CONTENTS

HEARING ON AGE DISCRIMINATION IN EMPLOYMENT ACT, PUBLIC SAFETY EXEMPTION

HEARING

BEFORE THE

SUBCOMMITTEE ON EMPLOYER-EMPLOYEE RELATIONS

OF THE

COMMITTEE ON ECONOMIC AND EDUCATIONAL OPPORTUNITIES HOUSE OF REPRESENTATIVES

ONE HUNDRED FOURTH CONGRESS

FIRST SESSION

HEARING HELD IN WASHINGTON, DC, JANUARY 24, 1995

Serial No. 104–3

Printed for the use of the Committee on Economic and Educational Opportunities

U.S. GOVERNMENT PRINTING OFFICE

88–349 CC WASHINGTON : 1995

For sale by the U.S. Government Printing Office
Superintendent of Documents, Congressional Sales Office, Washington, DC 20402
ISBN 0-16-046887-6

COMMITTEE ON ECONOMIC AND EDUCATIONAL OPPORTUNITIES

WILLIAM F. GOODLING, Pennsylvania, *Chairman*

THOMAS E. PETRI, Wisconsin
MARGE ROUKEMA, New Jersey
STEVE GUNDERSON, Wisconsin
HARRIS W. FAWELL, Illinois
CASS BALLENGER, North Carolina
BILL E. BARRETT, Nebraska
RANDY "DUKE" CUNNINGHAM, California
PETER HOEKSTRA, Michigan
HOWARD P. "BUCK" McKEON, California
MICHAEL N. CASTLE, Delaware
JAN MEYERS, Kansas
SAM JOHNSON, Texas
JAMES M. TALENT, Missouri
JAMES C. GREENWOOD, Pennsylvania
TIM HUTCHINSON, Arkansas
JOSEPH K. KNOLLENBERG, Michigan
FRANK D. RIGGS, California
LINDSEY O. GRAHAM, South Carolina
DAVE WELDON, Florida
DAVID FUNDERBURK, North Carolina
MARK SOUDER, Indiana
DAVID McINTOSH, Indiana
CHARLIE NORWOOD, Georgia

WILLIAM (BILL) CLAY, Missouri
GEORGE MILLER, California
DALE E. KILDEE, Michigan
PAT WILLIAMS, Montana
MATTHEW G. MARTINEZ, California
MAJOR R. OWENS, New York
THOMAS C. SAWYER, Ohio
DONALD M. PAYNE, New Jersey
PATSY T. MINK, Hawaii
ROBERT E. ANDREWS, New Jersey
JACK REED, Rhode Island
TIM ROEMER, Indiana
ELIOT L. ENGEL, New York
XAVIER BECERRA, California
ROBERT C. "BOBBY" SCOTT, Virginia
GENE GREEN, Texas
LYNN C. WOOLSEY, California
CARLOS A. ROMERO-BARCELÓ, Puerto Rico
MEL REYNOLDS, Illinois

JAY EAGEN, *Staff Director*
GAIL E. WEISS, *Minority Staff Director*

SUBCOMMITTEE ON EMPLOYER–EMPLOYEE RELATIONS

HARRIS W. FAWELL, Illinois, *Chairman*

THOMAS E. PETRI, Wisconsin
MARGE ROUKEMA, New Jersey
JAMES M. TALENT, Missouri
JAN MEYERS, Kansas
JOSEPH K. KNOLLENBERG, Michigan
DAVE WELDON, Florida
LINDSEY O. GRAHAM, South Carolina

MATTHEW G. MARTINEZ, California
MEL REYNOLDS, Illinois
DALE E. KILDEE, Michigan
PAT WILLIAMS, Montana
MAJOR R. OWENS, New York
DONALD M. PAYNE, New Jersey

HEARING ON AGE DISCRIMINATION IN EMPLOYMENT ACT, PUBLIC SAFETY EXEMPTION

TUESDAY, JANUARY 24, 1995

HOUSE OF REPRESENTATIVES, SUBCOMMITTEE ON EMPLOYER-EMPLOYEE RELATIONS OF THE COMMITTEE ON ECONOMIC AND EDUCATIONAL OPPORTUNITIES, *Washington, DC.*

The subcommittee met, pursuant to call, at 9:32 a.m., Room 2175, Rayburn House Office Building, Hon. Harris W. Fawell, Chairman, presiding.

Members present: Representatives Fawell, Petri, Roukema, Talent, Meyers, Weldon, Graham, Martinez, Reynolds, Williams, Owens, and Payne.

Also present: Representative Goodling [ex officio].

Staff present: Kathleen M. Gillespie, workplace policy counsel; Randel K. Johnson, workplace policy coordinator; Karen A. Wayson, staff assistant; and Lester Sweeting, Minority legislative associate.

Chairman FAWELL. Well, good morning to our witnesses and also to the Members of the subcommittee and to Mr. Martinez, who is the Ranking Minority Member of the Employer-Employee Relations Subcommittee, which is meeting here for the first time this year to consider the need for a public safety exemption under the Age Discrimination in Employment Act.

Just briefly, in reference to the rules under which we will be operating this year, although I may vary once in a while from this, only the Chairman and the Ranking Minority Party Member will be making oral opening statements, and Mr. Martinez and I might even skip that.

The concept that the committee as a whole is emphasizing is that we are here primarily to listen to you folks who are here to tell us something, and then to have the opportunity, under the five-minute rule, to propound questions to you and that will, I think, move it along just that much better.

In accordance with the committee rules, each Member is to be recognized in order of their appearance. I will presumably be the judge of that on our side of the aisle, and Mr. Martinez will take that heavy weight on his shoulders so far as the Minority is concerned to determine when his Members made their appearance here.

And so I will make a short statement, then ask the gentleman from California that he do the same, and then we will proceed with the witnesses.

The subject of today's hearing, the public safety exemption under the Age Discrimination in Employment Act, is familiar to the veterans of this committee, although not necessarily on this subcommittee. Legislation to extend the exemption allowing employers in the public safety arena to use maximum hiring and mandatory retirement ages, notwithstanding the general prohibition against age-based discrimination in the Age Discrimination in Employment Act, twice passed the House last Congress, once as a freestanding bill and once as an amendment to the crime bill. It is because the bill was bottled up in the other body that we address this issue once again.

Like many Members, I have found the issue somewhat troubling, as it requires us to balance two very worthy goals which, in this case, contradict one another. On the one hand, I strongly believe that the use of arbitrary age-based criteria in making employment decisions is a disservice to the skills and abilities that people bring to the work force. Especially as some of us wander up into the upper years, we tend perhaps to have that viewpoint more often than others.

On the other hand, one cannot very well ignore the issue of public safety and that currently there appears to be no better predictor of the ability in regard to firefighting and law enforcement officers than age—at least that is the feeling of some. I know that is not necessarily the feeling of all of the witnesses who will be testifying today.

The expiration of the public safety exemption in the Age Discrimination in Employment Act has left police and firefighting agencies in the same position that many businesses in a variety of industries find themselves, albeit without the same life or death consequences. In lieu of using age-based personnel policies, the public safety agencies have to develop physical and mental fitness tests that validly measure the ability of employees to perform their jobs and that will not be subject to an endless round of litigation.

As many employers know, that can be a Herculean task and the public safety community has very convincingly argued that such tests do not exist at the present time with respect to the fields of law enforcement and firefighting.

Unfortunately, we are in much the same position today that we were in 1986 to the extent that adequate guidance is not in place to enable police and fire departments to replace age-based personnel policies with physical and mental tests that will meet their work force needs while at the same time withstand judicial scrutiny.

I admit that I am no expert on the essential physical requirements of law enforcement or firefighting, but I find it persuasive that both management and labor in the public safety arena agree that fitness testing is not practical at this time.

While I read with interest the testimony of Captain Wall and I am certain there are other individuals in the field who can continue to perform at a later age, let us say, than most of us, I believe that the public safety exemption fairly walks the delicate line between protecting the employment rights of older Americans and ensuring the public safety needs of all Americans.

The Chair recognizes the gentleman from California.

Mr. MARTINEZ. Thank you, Mr. Chairman. That is hard to say, Mr. Chairman.

Chairman FAWELL. Yes, all of us have to get used to it.

Mr. MARTINEZ. Sitting on the other side, you have to get used to it, remember which door to go in.

I want to thank you, Mr. Chairman. I want to begin by welcoming the witnesses who have come a long way on very short notice and especially Captain Wall, who is a fellow Angeleno.

I appreciate the efforts you are making in holding this hearing. The hearing did not have the seven-day requisite notice, and I hope that is not going to set a precedent except in those situations, much like this, where there is an overwhelming need to do so.

As you know, Mr. Chairman, I support the aims of this legislation. In fact, I will agree to be a coauthor on the bill, but there are some reservations I have. You touched on it a little bit. There are exceptions.

Initially, when this legislation first came forward, I was one of those that, because I have seen, particularly, instances in my own district and my own city, where I was the mayor, exceptions, I thought you have got to make some accommodation for that exception. Captain Wall is here to argue that point.

I don't know how you do it. We have tried, and we tried last time and we ended up requiring a study to be done and we still haven't had, I don't think, definite results or proof from that study what to do, so I now believe that it is more important that we maintain the exemption and extend it for obvious reasons of suits that would arise because of people feeling they were done an injustice and still feel capable, without any way of making that assessment. I don't think we should leave it incumbent upon the agency that hires to be the one that has to prove otherwise.

So I say that the one reason that I am happy you are holding these hearings, too—because I notice and I see a couple of new faces on the other side, freshmen who are not privy like you and I to all the arguments on both sides of the issue, and I think it is important for them to hear it when they are going to vote on a very important matter like this. So I think that you are to be commended for that.

I would now like to relinquish the balance of my time so we might listen to the witnesses. After all, that is who we came to hear.

Chairman FAWELL. I thank the gentleman, and he is correct that the full amount of notice was not given the Minority. I appreciate the attitude of the Ranking Minority Member to proceed with this hearing nevertheless, which we all of course do wish to do.

Mr. MARTINEZ. Mr. Chairman, just one other thing. I would like to have my entire statement entered in the record.

Chairman FAWELL. Yes, without objection, it shall be.

[The prepared statement of Mr. Martinez follows:]

STATEMENT OF HON. MATTHEW G. MARTINEZ, A REPRESENTATIVE IN CONGRESS FROM THE STATE OF CALIFORNIA

Mr. Chairman:

I want to begin by welcoming our witnesses, some of whom have traveled a great distance on very short notice to be here today.

I would especially like to welcome my fellow Los Angeleno, Captain Frank Wall, to today's hearing.

Mr. Chairman, I appreciate your efforts in holding this hearing on the advisability of extending the sunset provisions in the Age Discrimination in Employment Act as the Act related to police and firefighters.

Mr. Chairman, I support the aims of the legislation which will again be considered in this Congress, and like you have heard most of the arguments on both sides of this issue.

And, although I am sure I've kept an open mind to any new and different information on either side of the issue, we have pretty much made up our minds as to the necessity for passing this legislation.

But I think we would both agree with a need for the newer Members of our committee to hear a certain amount of testimony before being asked to vote on this important matter.

Mr. Chairman, since this bill did pass the House in the 103d Congress, but died in the other body due to a lack of action, I can understand the need to be as expeditious as possible this year.

Therefore, I look forward to hearing our witnesses today and moving on.

Noting that most Members of Congress seem to agree that to push legislation to the floor with no hearings is not proper under our rules and shortchanges those who wish to learn more about the reasons for voting for or against the legislation, I commend you for calling this hearing.

That said, Mr. Chairman, I feel that I must remind you, on the record, that this hearing was called without the requisite seven day notice period as provided in the committee rules which the Majority party wrote earlier this month.

I trust that this is not going to set a precedent in the future absent a real compelling reason for ignoring the rules.

Again, let me say that I believe that it is good to start with a piece of legislation that does enjoy so much support on the committee, among all Members.

But let us listen to our witnesses with an open mind and give full attention to all points of view.

Thank you, Mr. Chairman.

Chairman FAWELL. At this time I think we will proceed, first of all, with Mr. Frederick H. Nesbitt, who is Director of the Department of Governmental Affairs and Political Action of the International Association of Fire Fighters, AFL–CIO and CLC in Washington. Following Mr. Nesbitt, the Honorable Patricia Ticer, Mayor of the City of Alexandria, who is not here as yet, but will be here shortly I understand and will be testifying. And then Mr. Robert Scully, Executive Director of the National Association of Police Organizations will follow with his testimony and then last, but certainly not least, Captain Frank Wall of the Los Angeles County Fire Department will be testifying.

So the Chair at this point recognizes Mr. Nesbitt.

STATEMENT OF FREDERICK H. NESBITT, DIRECTOR, DEPARTMENT OF GOVERNMENTAL AFFAIRS AND POLITICAL ACTION, INTERNATIONAL ASSOCIATION OF FIRE FIGHTERS, AFL–CIO–CLC, WASHINGTON, DC

Mr. NESBITT. Thank you, Mr. Chairman. My name is Fred Nesbitt. I am the Director of Governmental Affairs for the International Association of Fire Fighters, and I appear before you today on behalf of the Nation's 200,000 paid professional firefighters to express our enthusiastic support for your proposal to allow State and local governments to utilize mandatory retirement and maximum entry level ages for public safety officers.

As you stated in your opening statement—and you are well aware that in 1986 when Congress passed the amendments to the ADEA, they allowed for a seven-year temporary exemption for public safety officers while a study was conducted to see if there was a better way of determining age requirements for public safety officers other than using age-based criteria. Based on the results of the substitute of this committee last year, the House of Representatives voted unanimously to retain the ADEA public safety exemption. Unfortunately the legislation died in the other chamber when the Congress adjourned.

For several reasons which I wish to discuss today, we are asking you to move expeditiously to ensure that the permanent exemption from the ADEA is finally enacted in the 104th Congress.

The first point I want to make, Mr. Chairman, is to realize the unique nature of public safety occupations. The International Association of Fire Fighters opposes all forms of employment discrimination, and we are on record in opposition. However, we do believe that there are rare instances in which age is a valid criteria. Public safety work is one of those rare exceptions.

First, you must realize that the ability of a person to perform the work of a firefighter or police officer declines with age. The Penn State study recognized a direct correlation between advancing age and such vital factors as declining strength, reduced aerobic and anaerobic capacity and increased risk of sudden incapacitation. And the International Association of Fire Fighters—and we have collected data for many, many years in our membership—our own data show that older firefighters are much more likely to be killed on the fire ground or seriously injured in the line of duty than younger firefighters.

We are a unique group of people, Mr. Chairman. We are the only people who are taught to run into a burning building. Most people are taught to run out of a burning building, including Members of Congress, but we run into burning buildings; and when you take a look at the job of sworn officers who are firefighters, 95 percent of sworn officers are on the fire ground involved in fire suppression on a daily basis.

We don't have a bureaucracy. We don't have desk jobs, a lot of administrative positions. People who are sworn officers are performing the job of a firefighter; and when you talk about old firefighters, seasoned veterans of the department, you are talking about people who are in the 45–to–55 age category. These are the old firefighters, the seasoned people.

And what we find is, as age increases for a firefighter, the line-of-duty death possibilities increases substantially. The number one killer of firefighters in the line of duty is a heart attack. A heart attack is certainly related to 20 or 30 years of exposures to carbon monoxide, smoke, gases, hazardous materials; it is related to job stress, and it is certainly related to overexertion. And I think that makes the point that it is clearly correlated to age.

And when you take a look also at age you would imagine that younger firefighters would have more injuries simply because of inexperience, and maybe what you might call a "cowboy mentality" of hurry off and you fight a fire and be enthusiastic about it. But just the opposite is true.

As the age of a firefighter increases, the number of injuries increases, and the seriousness of the injuries increases. The time lost on the job increases substantially, and the likelihood of a disability pension increases dramatically. And it is basically related to age.

As the sponsor of the legislation said last year during the House floor debate, and we agree, Major Owens said, quote, "Age does indeed affect an individual's ability to perform the duties of a public safety officer. This is not a stereotype. This is not an ageism. This is a medical fact," end of quote.

And I think the other argument is that the workers themselves, the people who would be impacted and affected by this exemption, public safety occupations, are demanding and asking for the exemption from ADEA.

No issue in the union that I represent has broader or more passionate support from the membership than this particular issue. And I have included in my statement, Mr. Chairman, excerpts from letters from actual firefighters who have written to their Members of Congress. And we received copies of thousands of letters that they have written stating their support for such an exemption; and a couple of them I just want to cite quickly.

A firefighter from Virginia says, quote, "I assure you, when I retired at the age of 52, I was no longer able to fight fire or perform a heavy rescue. A 50-, 60- or 70-year-old person has no place in public safety."

Another firefighter from Massachusetts said, "I am asking you to support this amendment because I know that even at my relatively young age of 41, the best years of my physical capabilities are already behind me."

And lastly, a firefighter from New York said, "After 26 years as a firefighter and a supervisor, I can attest that the abuse that one's body takes from smoke, stress, physical exertion, injuries, and pain, both physical and mental, takes its toll."

Mr. Chairman, every organization representing rank and file firefighters, police officers, correction officers, has endorsed the public safety exemption from ADEA. And this also includes the employers, the State legislators, the cities, the counties, and the mayors. I don't know of any other issue in the public sector in which labor and management and employers are unanimous in their agreement and support of such an issue.

In short, there is not a single organization that speaks for the men and women in America's public safety occupations that opposes this exemption. And we believe that warrants serious consideration for this proposal.

Something that is very unique is the fact that in our occupation human lives are at stake. There is no other profession in which the inability to perform one's duties is so certain to result in the injury or death of innocent people—and it may be another firefighter, because they work in teams. In interior fire suppression, two firefighters are sent in together, and each one's life depends upon the ability of the other to perform the duties of a firefighter.

And, of course, also at risk are the very people that we are sworn to protect, the general public that we serve as sworn officers; and we believe the State and local governments must be given the flexibility to ensure a physically fit, fully qualified and balanced work force.

When you take a look at firefighters as a group, when they are sworn officers in the very beginning entering into the fire service, they are much healthier than the general population. We screen out disabilities. We screen out people with heart and lung problems because we know they couldn't do the job. So as a group of individuals, they are extremely healthy.

After 20 or 25 or 30 years on the job, we find that as a group, when they are ready to retire, they are substantially less healthy than the population at large; and it simply relates to the toughness of the job, the stress, the exertion, and the problems that they confront as a public safety officer.

I think another reason we need to look at this issue carefully is that the Federal Government has found that age is an essential part of structuring its own public safety force. We have mandatory retirement of Federal firefighters at age 55 and for law enforcement officers at age 57. The military has also found that a mandatory retirement age has proven essential to the maintenance of the national defense.

It is important to note that in the proposal that we are supporting we are not advocating a certain retirement age. We are not saying the retirement age should be 55 or 60 or 62 or 65. Rather, we are simply asking that local governments have the flexibility to set a retirement age based upon their needs in the same way that the Federal Government has had the flexibility to set a retirement age for their own firefighters and law enforcement officers.

One of the arguments used, and the question becomes, what about a fitness test? Isn't there something better than an age fitness test we could possibly use? Unfortunately, fitness tests are presently not a valid alternative to age.

First of all, fitness tests have a long history of being struck down by the courts because they are discriminatory. A test used in New England, a hose pull test for a firefighter, which sounds like a very good test, the court struck down because they said it was discriminatory against women because a hose pull test is a measure of upper body strength, and men have more upper body strength than women; therefore, if you use a hose pull test, you are ensuring that men will pass it and women will not.

And the ability of these tests to withstand legal challenges has increased significantly because of the Civil Rights Act of 1991, which prohibits norming, and because of the Americans with Disabilities Act, which places limitations on using fitness tests. To date, the EEOC has been unable to identify valid tests that meet the criteria that it is job related, nondiscriminatory, and that there are standards, standards both for the test itself and for the administering of the test.

Fitness tests tend to be invalid based upon the real-world experience of our membership. A test conducted in a doctor's office or a gymnasium does not replicate the physical demands of a firefighter on the fire grounds.

You can't measure for such things as extreme temperature fluctuations, smoke, stress, fatigue, and the speed at which a person may have to come out of a deep sleep and in minutes go into a full state of exertion. You can't measure the factor that someone may have been on the fire ground for 18 hours doing heavy, stressful work, have returned to the station for two minutes, and then must go out again and perform the exact same duties without any kind of rest, because our people work a 24-hour shift.

In the absence of such valid job performance tests, we believe that State and local governments must be given the option to utilize age as an employment criteria, and the key word is "option." We believe that State and local governments know best—based upon the demands of their department, based upon the kind of housing, the work force they have and so on and so forth—whether they should utilize age-based criteria such as an entry level age or a mandatory retirement age.

We have been pushing among our membership, and this is a goal of our union, to get away from testing and go to what we call a wellness program where a newly sworn officer who has already been tested physically fit for the fire service, what we try to do is counsel that person during his or her career of 20 or 25 or 30 years to make sure that they are physically fit, that they exercise, that they are concerned about their eating habits—in most departments, firefighters are not permitted to smoke anymore—and to make sure that their entire lifestyle, both on the job and off the job, enables them to be the most fit to do the job, to suffer the least amount of injuries, hopefully to avoid disability, and to ensure that after 20 or 25 years as a firefighter they can enjoy their retirement and they are not disabled or their lives are not shortened.

When you look at the Penn State Study recommendations, it is obvious that they predetermined that fitness tests were better predictors of ability than age. Dr. Landy, who conducted the study, selected studies which supported his conclusion, ignored those that refuted it, including data offered by the International Association of Fire Fighters, based upon our studies, and refused to consult as was mandated by the Congress with affected organizations. We believe that the research does not support the conclusions of the study.

Mr. Chairman, in conclusion, let me say that the public safety officers exemption to the ADEA has proven its value. Organizations representing rank and file public safety officers were divided on the question when Congress debated it in 1986, but in 1994 we were unanimous and in 1995 we are unanimous. We were in 1994, and we still are in 1995. The House of Representatives, which narrowly approved the exemption in 1986 after much heated debate, approved it with bipartisan support in the 103d Congress; and it was placed on the suspension calendar.

In light of the vast support for this proposal, we ask you to move as quickly as possible to give final approval to the public safety exemption to the Age Discrimination in Employment Act.

I thank you and will be happy to answer any questions you may have.

Chairman FAWELL. Fine. Thank you, Mr. Nesbitt.

[The prepared statement of Mr. Nesbitt follows:]

FREDERICK H. NESBITT

INTRODUCTION

Mr. Chairman, my name is Frederick Nesbitt and I am the Director of Governmental Affairs for the International Association of Fire Fighters. I appear before you today on behalf of the nation's 200,000 professional fire fighters to express our enthusiastic support for your proposal to allow state and local governments to utilize mandatory retirement and maximum entry ages for public safety officers.

Eight years ago, Congress enacted the Age Discrimination in Employment Act Amendments of 1986. The law exempted public safety occupations from the ADEA for a period of seven years while a study was conducted to determine whether such an exemption was warranted. Based on the results of that study, this committee and the House of Representatives voted unanimously in the last Congress to retain the ADEA public safety exemption. Unfortunately, the legislation never came to a vote in the other chamber and died when Congress adjourned.

For the reasons outlined below, the nation's fire fighters ask that you move expeditiously to ensure that a permanent exemption from the ADEA for public safety officers is finally enacted in the 104th Congress.

THE UNIQUE NATURE OF PUBLIC SAFETY OCCUPATIONS

Let me state from the outset that the International Association of Fire Fighters opposes all forms of employment discrimination and we fully support the Age Discrimination in Employment Act. It must be noted, however, that the ADEA was never intended to ban all uses of age in employment. There are rare instances in which age is a valid and irreplaceable criteria. Public safety work is one of those rare exceptions.

The first reason we believe that public safety is an exception to the general rule is based on numerous scientific studies that show the ability of a person to perform the work of a fire fighter or police officer declines with age. A close reading of the State study shows that advancing age is directly correlated to such vital factors as declining strength, reduced aerobic and anaerobic capacity, and increased risk of sudden incapacitation [see attached analysis of the Penn State Study conducted by the 103rd Congress' House Subcommittee on Select Education and Civil Rights]. Our own data show that older fire fighters are much more likely to be killed or seriously injured in the line of duty than younger ones. For example, while only one-seventh of all fire fighters are over age 50, they account for more than one-third of all fire fighter deaths.

As the sponsor of this legislation in the last Congress put it, *"Age does indeed affect an individual's ability to perform the duties of a public safety officer. This is not a stereotype. This is not ageism. This is a medical fact."*

A second reason that public safety is an exception to the general rule is that the intended beneficiaries of the law--the workers themselves--are demanding that public safety occupations be exempt from the ADEA.

Within in the IAFF, there is no issue which has broader or more passionate support among our membership. Resolutions supporting the exemption have been unanimously approved at several of our conventions. And we have received literally thousands of letters and phone calls from fire fighters across the nation expressing their support for this exemption. Allow me to share with you just a few excerpts from letters sent by our rank and file members to their Members of Congress on this issue.

> "When I went on the job two weeks before the maximum entry age of 36, I never thought I'd be supporting an act like this, but at age 58 I feel that it is really necessary to have both a maximum entry age and a mandatory retirement age."
>
> *Thomas Ginacchi*
> *Fairfield, CT*

> "As a professional fire fighter who could theoretically be discriminated against if exempt from the ADEA, I have no reservations in supporting the exemption."
>
> *Murray Orlov*
> *Columbus, OH*

> "In Chicago, we had a mandatory retirement age of 63. To work past age 63 in such a physical and demanding job is unsafe for those firefighters and the citizens they are supposed to protect."
>
> *Herbert Schmidt*

Chicago, IL

"As a 53 year old fire fighter, I know that my physical capabilities have diminished considerably. The physical demands of our job at a fire scene with life at risk are as great as any professional athlete. The difference is that that if we are unable to perform our duties, human life may be lost, not the game."

Ken Van Dyke
Grand Rapids, MI

"I have never written you before, but I had to this time because I feel this is an important issue not only for myself, but also for the safety of all the public."

Alex Horvath
Brook Park, OH

"The City of Philadelphia, like many other municipalities throughout the nation, is facing tough financial times. Philadelphia has decided to remediate that difference on the shoulders of its fire fighters. There are fewer fire companies and less personnel on each apparatus. To remove the maximum entry levels and mandatory retirement age would further jeopardize fire fighter safety, as well as the lives of those whom we serve."

James Kerrigan
Philadelphia, PA

"I assure you when I retired at age 52 I was no longer able to fight fire or perform a heavy rescue....A 50, 60, or 70 year old person has no place in public safety."

Buck Swartz
Alexandria, VA

"Firefighters and police officers must work as a team. We depend on the other members of our crew to have the strength and savvy to save our life if the need arises. If we are unable to do our job, people die."

Carl Below
Eugene, OR

"I am asking you to support this amendment because I kow that even at my relatively young age of 41, the best years of my physical capabilities are already behind me....I cannot believe that any one using a common sense approach would believe that 70 or 80 year old fire fighters would be a good thing for their communities well being."

Robert Johnson
Harwich Port, MA

"To allow entry level positions to be open to all, regardless of age, could also have a devastating effect on our goal, to protect the public we serve from natural and man-made disasters that we face in everyday life.

Frank Nefos
Birdsboro, PA

"Fire fighting is at best, a very strenuous activity. We go from a semi-state of calm to full exertion within a matter of moments. The ability to do this effectively is a matter of life or death...."

Robert Lee Bragg
Charlottesville, VA

"After 26 years as a firefighter and supervisor I can attest [that] the abuse that one's body takes from smoke, stress, physical exertion, injuries and pain (both physical and mental) takes its toll. As an old Firehouse saying goes, 'It's a young man's job.' If a firefighter is unable to do his job, there is not a delay in the work being done, but maybe death."

Daniel Finegan
Chester, NY

Significantly, fire fighters are not alone within the public safety community in support of this proposal. Every organization representing rank and file fire fighters, police officers and corrections officers has endorsed the public safety exemption from the ADEA. Even the AFL-CIO, the nation's foremost voice for working people and long time champion of employment discrimination legislation, has called for exempting public safety occupations from this law.

In short, there is not a single organization that speaks for the men and women in America's public safety occupations that opposes this exemption. We believe this fact alone warrants serious consideration of the proposal.

But the most important reason that public safety occupations are an exception to the general rule against age-based employment criteria is simply that human lives are at stake. There is no other profession in which the inability to perform one's duties is so certain to result in the injury or death of innocent people. This price is too high to take chances. States and local governments must be given the flexibility to ensure a physically fit, fully qualified and balanced workforce.

THE FEDERAL GOVERNMENT EXPERIENCE

The successful experience of the federal government in using age-based criteria provides an excellent example of how entry and exit ages can help build a balanced, efficient and effective workforce.

Even though Congress applied the ADEA to the federal sector in 1974, Congress has also retained the mandatory retirement age of 55 for federal fire fighters and law enforcement officers. The federal government has found that age is an essential part of structuring its own public safety plans. Granting the federal government this leeway without granting state and local governments the same option is patently unfair.

Perhaps even more relevant is the use of mandatory retirement ages in the military. Congress has historically treated the armed forces different from other employers, but fire fighting is in many ways closer to the military than it is to other types of work. We are America's domestic defenders, and the nature of our work and our method of operation parallels the military in several respects.

The military has found that its use of a mandatory retirement age has proven essential to the maintenance of our national defense. In recent years there has been a great deal of debate surrounding the rationale for various military requirements. Nevertheless, to our knowledge no one has ever questioned the use of military retirement ages.

It is important to note that we are not advocating the use of a nationwide mandatory retirement age. Rather, we are simply asking that local governments have the same flexibility in seeking to ensure public safety that the federal government has with its fire fighters and law enforcement officers.

FITNESS TESTS

Those who oppose the use of mandatory retirement ages in public safety occupations point to fitness tests as a better way to ensure a qualified workforce. Unfortunately, fitness tests are presently not a valid alternative to age. Allow me to explain a few problems we currently have with fitness tests and how we hope to address these shortcomings.

First, fitness tests have a long history of being struck down by courts because they are discriminatory. For example, the U.S. District Court for the District of New Hampshire recently ruled that a hose-pull test given to fire fighter applicants is invalid because it discriminates against women. After hearing evidence that the test requires upper body strength not present in the average woman, the Court enjoined the city from using the test and ordered the city to hire a female applicant who had failed the test.

Mr. Chairman, I find it especially significant that the specific test that was rejected by the court as discriminatory in this case is a hose-pull test. This test is clearly job-related. Pulling a hose is something every fire fighter must do. Nevertheless, the court threw out the test based solely on the fact that males find it easier to pass the test than females.

The ability of fitness tests to withstand legal challenges has become increasingly difficult due to the enactment of two recent civil rights laws. The first, the Civil Rights Act of 1991, bans the use of test "norming." Norming has been widely used in the fire service as a way to ensure more women qualify for positions, and the inability to norm scores will make it much more difficult to design a test which is both meaningful and doesn't discriminate based on gender.

The second civil rights statute that will affect fitness testing is the Americans with Disabilities Act. The ADA contains broad new limitations on the use of fitness testing which make it far more difficult to defend their use in court.

The advent of these two laws, in conjunction with Title VII of the Civil Rights Act of 1964, will make it extraordinarily difficult and perhaps impossible for local governments to design meaningful physical fitness tests which can ensure a qualified workforce. To date, EEOC has been unable to identify valid tests that will meet these criteria.

But the more significant reason why we believe fitness tests to be invalid is based on the real world experience of our membership. Tests conducted in a doctor's office or in a gymnasium simply do not replicate the physical demands placed on fire fighters at a fire ground operation. We are aware of no fitness tests that can account for such factors as extreme temperature fluctuations, smoke, stress, fatigue and the speed at which a person must go from a sedentary state to a state of full exertion. A fitness test can measure a person's ability to climb three flights of stairs carrying a hose. But it can not measure a person's ability to perform that task in a smoke filled building in sub-zero temperatures only minutes after being awoken from a deep sleep.

In short, we believe that performance tests are not valid predictors of an individual's ability to perform the job of a fire fighter. In the absence of valid performance tests, there currently is no better predictor of ability than age.

Mr. Chairman, allow me to stress that we are not opposed to the concept of testing. We support language included in your legislation which will continue the search for valid performance tests. Under this provision, the Equal Employment Opportunity Commission is required to study the extent to which performance tests can accurately measure the abilities necessary to be a public safety officer, taking into account such issues as federal anti-discrimination laws.

Nevertheless, the current reality is that valid performance tests that can replace mandatory retirement ages have not yet been identified. In the absence of such valid performance tests, states and local governments must be given the option to utilize age as an employment criteria.

Before leaving the issue of physical fitness testing, I would like for one moment to address the Penn State's study recommendations regarding these tests. We are of course aware that Dr. Frank Landy concluded in his study that fitness tests are reliable predictors of ability, and I want to share with you why we believe these findings should be taken with a large grain of salt.

From the very beginning of Dr. Landy's study, it was clear to us that he had pre-determined that fitness tests were better predictors of ability than age, and all of his research was geared toward supporting that conclusion. Dr. Landy selected studies which supported his conclusions, ignored those that refuted it, and flagrantly violated the provisions of the law which required him to consult with affected organizations. Even with these extraordinary efforts to skew the results of his study, it is dubious at best that his research supports his conclusions.

Although we realized early on that Dr. Landy had made up his mind before the study began, we never before understood the reason for his bias. We now do.

When not conducting studies or teaching, Dr. Landy designs physical fitness tests for police and fire departments. He is in a position to receive a significant personal economic benefit from the abolition of mandatory age limits. This serious conflict of interest should have prevented him from receiving the contract to conduct this study in the first place. At the very least, members of this Committee need to consider this conflict in assessing the relevance of the Penn State Study.

If you wanted to know if buying a used car is a good idea, the last person you would ask for an unbiased opinion is a used car salesman. In the same light, Dr. Landy's bias must be taken into account when reading the conclusions of his work.

CONCLUSION

Mr. Chairman, the Public Safety Officers exemption to the ADEA has proven its value. Organizations representing rank and file public safety officers were divided on the question when Congress debated it in 1986, but are now unanimous in their support. The House of Representatives, which narrowly approved the exemption in 1986 after heated debate, adopted it with bi-partisan support in the 103rd Congress on the suspension calender.

Unfortunately, the 1986 exemption expired on January1 of last year, and public safety agencies are now in a state of limbo. In light of the vast support for this proposal, we ask you to move as quickly as possible to give final approval to the Public Safety Officers Amendment to the ADEA.

Thank you. I would be happy to answer any questions you may have.

Chairman FAWELL. Mr. Scully, we will now receive your testimony.

STATEMENT OF ROBERT SCULLY, EXECUTIVE DIRECTOR, NATIONAL ASSOCIATION OF POLICE ORGANIZATIONS, WASHINGTON, DC

Mr. SCULLY. Mr. Chairman, thank you. My name is Bob Scully. I am the Executive Director of the National Association of Police Organizations. NAPO represents over 175,000 sworn law enforcement officers throughout the country. We represent over 3,000 police associations or unions throughout the country.

Mr. Chairman, before becoming the Executive Director of NAPO, I served 10 years as its elected national President and in 1993, I retired from the Detroit Police Department after 26 years of service. I was indeed a practitioner on my job in Detroit.

The last 20 years I also had responsibility as the elected Vice President of the Detroit Police Officers' Association to represent the 3,000 sworn police officers in the City of Detroit. I represented them in the negotiation process, I handled discipline matters for them, and I also handled disability and pension matters for them.

Hiring and retirement ages are not an arbitrary figure. These ages that were established in the City of Detroit and the City of New York, in the City of Los Angeles—in the cities across this country, in counties across this country and in State governments across this country—were arrived at after careful consultation between employee representatives and employer representatives, after consultation with medical section representatives. These are doctors who are hired by police departments and fire departments to care for officers who are sick or injured, and after consultation with actuaries.

The two most important items, I feel, that were taken into consideration, because I was part of that process over the last 20 years, were, number one, the issue of safety—safety of the employee, safety of the employee's coworker and the safety of the American public that we are sworn to protect. The second most important item that we took into consideration was the financial integrity of the pension system that we all hope to collect a benefit from some day.

I would like to give you an example. Recently—and I stress "recently"—in the City of Los Angeles, they have graduated a recruit out of their academy who is 55 years of age. I think we have to look at two situations here. It is the safety situation and it is the financial stability of the pension system.

The likelihood of an individual completing the necessary years of service to collect a normal retirement, when hired at 55 years of age, is slim to none if that officer is required to do the normal work of a law enforcement officer. In all likelihood, that officer within 10 years will have an injury that will not permit him to complete his tour of duty. So you will have an officer who has contributed to a pension system for 10 years—and if you follow the mortality tables today, there is a likelihood that that individual may live until he is 84 years old—who will collect from that pension system many more years than he contributed; and odds are, the individual will be collecting a pension that is 75–to–100 percent pay because of his duty disability and not the normal 50 percent at his retirement. It will be a financial drain on the pension system.

And the safety issue also comes into consideration here because as law enforcement officers, as Fred said in the fire field, we work as teams. We rely on backup. The general public relies on us to respond in a timely fashion, and I really think the safety issue comes into play.

Pre–1994, the maximum entry level age in the City of Detroit was 35 years of age. The retirement eligibility was after 25 years of service, and the mandatory retirement age was 60 years of age, and an individual could apply for two two-year extensions, but at age 64, they had to leave. We need this type of mutually agreed-upon age-based retirement system to continue. We need Congress to make permanent the public safety exemption to the Age Discrimination in Employment Act.

Law enforcement officers have the most dangerous profession in America today. According to the latest FBI statistics that were just released, 157 law enforcement officers were killed in the line of duty in 1994. There were another 150 police officers who are alive today only because they were wearing bulletproof vests when they were shot in 1994; the total could well be over 300 killed. The FBI reports that there were 66,000 assaults on police officers resulting in over 24,000 serious injuries to police officers in 1994.

The job of a law enforcement officer is both physically demanding and emotionally draining. Law enforcement is, by its very nature, a young person's job. And I am asking the Members of this subcommittee and eventually the Members of the full committee and the Members of this House to do what it did in the last Congress to support public safety and pass the public safety exemption to the Age Discrimination in Employment Act. I am sure that if Members of this committee searched around in your local communities, you could probably find a police officer or a firefighter to come forward and say, I disagree with the union that represents me. I think you could find that in any occupation, and I know you could find it in police.

But let me tell you that I am also a member of AARP, and there are probably a few other people in this room who are also members of AARP, and we don't necessarily agree with their position on this isolated item right here. But the majority, the vast majority of police officers in this country, every rank and file organization, every management organization, is unanimous in its support for the passage of this very important amendment.

Mr. Chairman, I would ask that my written and oral remarks be made part of the record. Thank you very much, and I am open to questioning.

Chairman FAWELL. And thank you. And without objection, certainly your full statement will be incorporated into the record.

[The prepared statement of Mr. Scully follows:]

ROBERT T. SCULLY

Good morning, Chairman Goodling, members of the Subcommittee, it is a distinct pleasure to have the opportunity to appear before you today as this panel considers legislation to reinstate the exemption for public safety employees in the Age Discrimination in Employment Act.

My name is Robert Scully and I am the Executive Director of the National Association of Police Organizations (NAPO) representing over 175,000 sworn law enforcement officers and 3,000 police unions and associations throughout the country. I am also the former President of NAPO and Vice President of the Detroit Police Officers Association and I recently retired from the Detroit Police Department after 26 years of service.

In 1983, Congress enacted the Age Discrimination in Employment Act amendments which exempted public safety employees from the ADEA for a period of seven years while a study was conducted to determine whether such an exemption was warranted. The exemption allowed police and fire departments to continue use of mandatory retirement and maximum entry ages if in place as of March, 1983.

The study, conducted by Penn State University, concluded that although age is a determinant of a person's ability to perform the functions of a public safety officer, there is no exact age at which a person can be said to be unqualified. NAPO agrees with this statement, but it does not automatically follow that age-based retirement ages do not work. The study contains serious flaws. An example is that it recommended the use of fitness tests as an alternative to age-based retirement and then failed to identify any specific physical or mental fitness tests or even any standards to be satisfied by such tests.

NAPO strongly supports an amendment to the Age Discrimination in Employment Act which would allow state and local public safety agencies to utilize mandatory retirement and maximum entry ages because age provides a more reliable indicator of physical ability than physical and mental fitness tests. The age-based criteria also protects public safety and the safety of other officers. In addition, the resintatement of the ADEA exemption will enable police departments to avoid costly litigation with widely disparate results.

While there are many reasons for the continuation of mandatory retirement ages for law enforcement officers, we believe the most compelling is to safeguard the public safety. Because officers perform tasks that require great physical strength and high levels of stress, they must be in reliable physical condition. There are serious, sometimes fatal, consequences when police officers fail to properly respond and control incidents. Other consequences could include serious physical injury and property damage.

In policing, we must always be prepared for the worst case scenario and have sufficient numbers of officers qualified, ready and able to meet the challenges that arise daily.

Older public safety officers are more likely to suffer catastrophic medical events thus jeopardizing the safety of their fellow officers, as well as increasing more expensive duty disability retirements. As older people lose essential, job-related skills over time, the safety of fellow officers and the public at large may be placed at risk. It is not realistic to expect 65 year old officers to expend the physical energy necessary to subdue violent criminals.

Although, there are numerous positions that senior citizens can hold well past the age of 65, policing is not among them.

Even the Penn State study found that if state and local public safety agencies did not use age-based criteria, an estimated 8.75 per 100,000 incidents per year would occur in which people die or are injured as a result of the sudden incapacitation of a public safety officer. I doubt the public would find this an acceptable level to obtain an appearance of non-discrimination. I firmly believe that protecting the public safety justifies this exemption.

Law enforcement is by its very nature a young person's job. Most public safety retirement systems are funded based on years ranging from 20 - 30 years. The expiration of the ADEA exemption will jeopardize the integrity of many retirement systems through increased claims for duty disability retirements. This will also give state legislators, community commissioners and city mangers the incentive to increase the years of service and age requirements on the pension systems as we currently know them.

In conclusion, if police officers are unable to perform their job, then the public will suffer and people will die. NAPO feels that it is critical that Congress enact the exemption to the ADEA immediately to allow public safety agencies to continue to use mandatory retirement and maximum entry ages.

Thank you again Chairman Goodling for the opportunity to share our views with you and I look forward to working with you on this important public safety legislation.

Chairman FAWELL. Now, Captain Franklin A. Wall, Jr. I had before me a statement that you were representing the Los Alamitos——

Mr. WALL. That is my residence. I live in Los Alamitos, California. I work for the County of Los Angeles.

Chairman FAWELL. I have had the opportunity of reading your statement. You indeed are the loyal defense here. I hope you don't feel outnumbered, but please proceed at this time, expressing your view both personally and also in representing the American Association of Retired Persons.

STATEMENT OF CAPTAIN FRANKLIN A. WALL, JR., LOS ANGELES COUNTY FIRE DEPARTMENT, ON BEHALF OF THE AMERICAN ASSOCIATION OF RETIRED PERSONS

Mr. WALL. Okay.

Mr. Chairman, I am Frank Wall, and I am a captain with the Los Angeles County Fire Department. I am not wearing my uniform because I am here to discuss my job and represent AARP, not my department or county. And this is quite an act to follow.

I am 60 years old and in peak physical and mental condition. If forced retirement is reinstated, I will be thrown off of my job. It is a job that I love and at which I excel.

For most of my life, I have been involved with firefighting. My father was an engineer with the City of Los Angeles Fire Department for 33 years. He voluntarily retired at age 65, still fit and healthy; and he is 84 years old today and still drawing his pension.

I have wanted to be a firefighter since I was seven years old. I was a Marine and did a plumbing apprenticeship. But in 1960, I took a 50 percent cut in pay to join the County of Los Angeles Fire Department. I have worked in almost every firefighting operation, the Fire Suppression Bureau, the Fire Prevention Bureau, the Services Bureau; and I have been to recruit training camp. I have worked most of my time in the most active stations in L.A., mainly South Central Los Angeles. I have worked with firefighters of all ages and abilities in our department and other departments.

Currently, I am in charge of a fire suppression camp in the Santa Monica Mountains near Malibu. I supervise fire crews on wildland fires in rugged terrain, which requires extreme physical exertion. I also supervise crews on other disasters, such as earthquakes, floods, air rescue operations and mountain rescues. I coordinate all the daily activities, all of the out-of-camp construction at other fire facilities, and am responsible for the ongoing training of our crew and the rookie crew members, plus all the paperwork.

The job requires physical strength and stamina. In the 1988 Yellowstone fire storms, we worked 12-hour shifts for a month straight with the U.S. Marine Corps. Equally important, it requires leadership, experience, and calm in the face of danger.

Keeping fit is an absolute necessity in our fire department. Periodic medical exams have been required for all firefighters of all ages and in all jobs since the early 1970s. We also have a mandatory physical fitness program which consists of 1½ hours of rigorous cardiovascular exercise every day we are on duty. It is my responsibility to make sure that our crew of 24 works out and passes the test.

I don't know how much it costs to give these tests, but whatever it is, it is worth it, both in terms of public safety and probably financially also. We have fewer injuries and disabled people. There is no question that the firefighters in my department today are in much better shape than when I joined the force.

I have never failed a fitness test, and for the past 10 years, I have taken one every year. Like many of my colleagues, I have developed a personal fitness program that keeps me in top shape, so I know that I will pass the test. I run or hike a minimum of 12 miles a week, work out in the weight room and do strength exercises for my legs and upper body. I have logged over 10,000 miles of running since 1971 when our program was initiated. I have completed two marathons and many half-marathons and 10K runs. I didn't do any of this before the fitness requirement was instituted.

About a year ago, I worked 19 days straight fighting the enormous Greenmeadow and Old Topanga fires in Malibu. You may have seen these on the TV news. My assignments required every single bit of training and experience I have accumulated in my 35-plus years on the job. One night my strike team, which was expanded to include 42 State inmates, was assigned to a crew working to save the Foundation for the Junior Blind Camp and Retreat. I sized up the situation, devised a plan and explained it to our 45 people and went to work setting a back-fire. This was tough work that had to be done quickly. We cut through dense brush and starting firing out, which was basically setting fires between the Junior Blind structures and the main fire to consume fuel that would otherwise feed the main fire.

As supervisor, I coordinated all the work of the three crews, had to continuously adjust the work plan to changing wind conditions—the winds were gusting to 60 miles an hour—watch for spot fires and immediately jump on those and make sure everything was done quickly. And we saved the Junior Blind Foundation.

After the recent Northridge earthquake, I worked 12 days straight with our strike team cleaning up and demolishing damaged structures, laying temporary water maintenance and providing relief services directly to the public.

Firefighting doesn't work if you don't trust your team leader or the person next to you. Although I keep up with them in every task, our crew really looks to me for my experience and judgment. The outstanding job they do on all emergencies proves that they trust me and that I am doing a good job for the people of Los Angeles County.

I am not unique in either experience, age or fitness. In the summer of 1993, we worked on a mutual aid fire in the Griffith Park area within the City of Los Angeles. The fire was burning in steep, rugged terrain heading for the park's observatory. A Los Angeles City engine company crew, led by Captain John Peterson, was laying hose lines to support our crews with water. They did an outstanding job under very demanding and complicated physical conditions.

Captain Peterson is 68 years of age. Neither I nor our team would hesitate to work with him again. L.A. City Fire Department is a class one, nationally known fire department with no mandatory age requirement; and they do have a periodical medical and physical fitness program.

I have been a member of the International Association of Fire Fighters since I joined the Department. I praise them for their efforts to make the fire service safer and better for firefighters and for working to get us good benefits and job conditions.

But they are wrong on this one and many of my fellow firefighters agree with me.

The International said that it is not possible to devise good fitness tests. That is just not true. We do it, L.A. City does it and lots of other departments do it. I can't understand why any department wouldn't do it. Our fitness testing and fitness programs have encouraged our firefighters to stay in shape, eat better, and stop smoking and drinking. I would rather work with any one of the people in my department or L.A. City than someone from a department that doesn't test like we do.

Firing competent and skilled firefighters because of their age is blatant discrimination and bad for the public. The sole measure of the public safety should be whether the firefighter can do the job.

Mr. Chairman, I am not a politician. I am out on the line every day; I know what it takes to do the job. I have been in the job of protecting the public for over 35 years. If I ever felt that I wasn't up to it, I would quit. Certainly my crew would tell me to quit, because their lives are in my hands. I have seen firefighters who have been asked to leave because they couldn't do the job or pass the fitness test.

In the year since mandatory retirement for firefighters was eliminated, there have been no horror stories of older firefighters failing to do their jobs.

I am not working for the money. My pension is equal to my salary, so if I retired, I would have the same money coming in that I am making right now; but I know this isn't the time for me to leave and my colleagues know it, too.

I hope you will agree and not vote for me to lose my job. Thank you.

Chairman FAWELL. All right. Thank you, Captain Wall; I appreciate your testimony.

[The prepared statement of Captain Wall follows:]

FRANKLIN A. WALL, JR.
Captain, Los Angeles County Fire Department

Mr. Chairman, I am Frank Wall, a Captain with the Los Angeles County Fire Department. I'm not wearing my uniform because I'm here to discuss my job and represent AARP, not my department or County.

I am sixty-one and in peak physical and mental condition. If forced retirement is reinstated, I will be thrown off my job. It's a job that I love and at which I excel.

For most of my life I've been involved with firefighting. My father was an Engineer with the City of Los Angeles Fire Department for 33 years. He voluntarily retired at age 65, still fit and healthy. I've wanted to be a firefighter since I was seven years old. I was a Marine, and then did a plumbing apprenticeship. But, in 1960, I took a 50% cut in pay to join the County of Los Angeles Fire Dept. I've worked in almost every firefighting operation -- Fire Suppression Operations, Fire Prevention Bureau, Service Bureau and as a Recruit Training Captain. I've worked most of my time in the most active fire stations in Los Angeles, mainly South Central LA. I've worked with firefighters of all ages and abilities in our department and others.

Currently, I'm in charge of a Fire Suppression Camp in the Santa Monica Mountains near Malibu. I supervise fire crews on wildland fires in rugged terrain, which requires extreme physical exertion. I also supervise crews on other disasters such as earthquakes, floods, air rescue operations, and mountain rescues. I coordinate all daily activities, all of the out-of-camp construction at other fire facilities, and am responsible for the ongoing training of our fire crews and rookie crew members, plus all the paper work. The job requires physical strength and stamina -- in the 1988 Yellowstone Firestorms we worked 12-hour shifts for a month with the U.S. Marine Corps. Equally important, it requires leadership, experience, and calm in the face of danger.

Keeping fit is an absolute necessity in our fire department. Periodical medical exams have been required for all firefighters -- of all ages and in all jobs -- since the early 1970's. We also have a mandatory physical fitness program, which consists of 1 1/2 hours of rigorous cardiovascular exercise every day you are on duty. It's my responsibility to make sure my crew of 24 works out and passes the tests. I don't know how much it costs to give the tests, but whatever it is, it's worth it both in terms of public safety and probably financially also. We have fewer injuries and disabled people. There's no question that the firefighters in my department today are in much better shape than when I joined the force.

I've never failed a fitness test, and for the past ten years I've taken one every year. Like many of my colleagues, I have developed a personal fitness program that keeps me in top shape so that I know I'll pass the test. I run or hike a minimum of 12 miles a week, work out in the weight room and do strength exercises for legs and upper body. I've logged over 10,000 miles of running since 1971, completed two marathons and many half-marathons and 10K runs. I didn't do <u>any</u> of this before the fitness requirement was instituted.

About a year ago, I worked nineteen days straight fighting the enormous Greenmeadow and Malibu fires. You may have seen these on the TV news. My assignments required every single bit of training and experience I have accumulated in my 34-plus years on the job. One night, my strike team -- which was expanded to include 42 state inmates -- was assigned to a crew working to save the Foundation for the Junior Blind camp and retreat. I sized up the situation, devised a plan and explained it to my 45 people, and went to work setting a back-fire. This was tough work that had to be done quickly. We cut through dense brush and started "firing out," which was basically setting fires between the Junior Blind structure and the main fire to consume fuel that would otherwise feed the main fire. As supervisor, I coordinated all the work of the three crews, had to continuously adjust the work plan to changing wind conditions -- the winds were gusting to 60 mph -- watch for spot fires and immediately jump on them, and make sure everything was done quickly. We saved the Junior Blind Foundation.

After the recent Los Angeles earthquake in, I worked twelve days straight with my strike team cleaning up and demolishing damaged structures, laying water lines and providing relief services directly to the public.

Firefighting doesn't work if you don't trust your team leader, or the person next to you. Although I keep up with them in every task, my crew really looks to me for my experience and judgment. The outstanding job they do on all emergencies proves that they trust me and that I'm doing a good job for the people of Los Angeles County.

I am not unique in either experience, age, or fitness. In the summer of 1993 I worked on a mutual aid fire in the Griffith Park area, within the City of Los Angeles. The fire was burning in steep, rugged terrain heading for the Park's observatory. A Los Angeles City engine company crew, lead by Captain John Peterson, was laying hose lines to support our crews with water. They did an outstanding job under very demanding and complicated physical conditions. Captain Peterson is 68. Neither I nor my team would hesitate to work with him again. LA City Fire Department is a Class I, nationally known fire department with no mandatory age requirement and a physical fitness program.

I have been a member of the International Assn. of Firefighters since I joined the Department. I praise them for their efforts to make the fire service safer and better for firefighters, and for working to get us good benefits and job conditions. But they are wrong on this one and many of my fellow firefighters agree with me.

IAFF says that it's not possible to devise good fitness tests. That's just not true. We do it, LA City does it, and lots of other departments do it. I can't understand why any department wouldn't do it. Our fitness testing and fitness programs have encouraged our firefighters to stay in shape, eat better and stop smoking and drinking. I'd rather work with any one of the people in my department or LA City than someone from a department that doesn't test like we do.

Firing competent and skilled firefighters because of their age is blatant discrimination and bad for the public, too. The sole measure of public safety should be whether the firefighter can do the job.

Mr. Chairman, I'm not a politician. I'm out on the line every day. I know what it takes to do the job. I've been in the job of protecting the public my entire life. If I ever felt that I wasn't up to it, I'd quit. Certainly, my team would tell me to quit, because their lives are in my hands. I've seen firefighters who have been asked to leave because they couldn't do the job or pass the fitness test.

In the year since mandatory retirement for firefighters was eliminated, there have been no horror stories of older firefighters failing at their jobs.

I'm not working for the money. My pension is almost equal to my salary, so if I retired I'd have the same money coming in every day as I do now. But I know this isn't the time for me to leave, and my department and my colleagues know it too. I hope you will agree and not vote for me to lose my job.

88-349 O - 95 - 2

Chairman FAWELL. I am happy to see the Honorable Patricia Ticer—is that the correct pronunciation?

Ms. TICER. It certainly is, yes, Mr. Chairman.

Chairman FAWELL. Thank you for joining us——

Ms. TICER. Thank you very much.

Chairman FAWELL. [continuing] mayor of the City of the Alexandria. So if you are ready to proceed——

STATEMENT OF HON. PATRICIA TICER, MAYOR OF THE CITY OF ALEXANDRIA, VIRGINIA, ON BEHALF OF THE NATIONAL LEAGUE OF CITIES, WASHINGTON, DC

Ms. TICER. I am ready. Thank you. I apologize for being late, but I was at another coalition meeting in order to try to save the Economic Development Administration earlier this morning.

Good morning. My name is Patsy Ticer. I am the Mayor of Alexandria, and I am here for the National League of Cities on behalf of 150,000 municipal elected officials and municipal organizations in 49 States represented by the League of Cities. The National League of Cities appreciates the subcommittee's holding this timely hearing to consider a subject of importance to cities and towns across the country and to the law enforcement officers and firefighters who provide essential public safety services to these communities.

Our purpose in presenting testimony is to urge the subcommittee to approve legislation that would allow all cities and towns the option—just the option—to use age-based criteria for hiring and retirement of public safety personnel. Further, we would urge Members to include language that would relieve municipalities from liability resulting from the expiration of the ADEA exemption on December 31, 1993.

As the subcommittee is aware, the seven-year exemption of police and fire personnel from the Age Discrimination in Employment Act, as it applies to hiring and discharge, only expired December 31, 1993. This means that cities and towns have been left unable to enforce age-based hiring and retiring decisions.

Over the past several years, NLC has worked hard in partnership with national organizations representing rank and file law enforcement officers and firefighters, labor and management together, to ensure that all cities and towns have the right to determine for themselves the types of hiring and retirement policies that meet the needs of each community. Such policies are essential for elected officials to fulfill their legal responsibility to provide for the public's safety.

Last year, H.R. 2722 was approved by the House of Representatives without controversy. The Senate refused to take up the measure. The effects of this failure have resulted in a serious disruption in long-standing employment policies and practices for those cities included in the exemption. Those cities and towns not covered by the exemption, those that did not have age-based policies on the books prior to 1986, have been forced to implement policies that do not represent the best interests of the residents of their respective communities.

Our request is consistent with current Federal policies and practices with respect to Federal public safety personnel. In 1974, as you know, the Congress applied the ADEA to the Federal sector. However, Congress retained the mandatory retirement age of 55 for Federal firefighters and law enforcement personnel. NLC believes the exemption for Federal public safety agencies from the ADEA requirements is fair; the same flexibility should apply to all levels of government.

Mr. Chairman, my remarks will focus on six specific issues that will explain why NLC supports legislation that would make permanent the exemption and expand it to all municipalities.

First, and perhaps most importantly, the bulk of city governments and units of organized labor that represent the law enforcement and firefighting personnel working for these communities support legislation that would extend to all cities and towns the option to implement age-based criteria.

Second, Federal law provides other special recognitions of retirement ages for law enforcement and firefighting personnel.

Third, public safety employment concerns raised during the enactment of the limited exemption in 1986 remain important.

Fourth, the reports, studies and guidelines that were to have been done by the Department of Labor and EEOC well in advance of the exemption's expiration in 1993 have not been completed.

Fifth, the absence of well-developed validated and tested alternatives is resulting in costly and needless litigation.

Sixth, the impact on affirmative action goals, which are critical to serving our ever-more culturally diverse communities, is of great concern to municipal leaders.

On the issue of extending the previous exemption to include all cities and towns, there is an agreement between city governments and major segments of organized labor representing law enforcement and firefighting personnel which these city and State governments employ.

As you are well aware, this level of consensus does not always exist. NLC would urge the subcommittee to take advantage of a win-win situation by advancing legislation to exempt cities and towns from the ADEA.

Two, Federal law in other places provides special recognition of the particular structure of public safety occupations so that an exemption would not be inconsistent with other statements of Federal policy.

For example, section 415 of the Internal Revenue Code, which governs pension issues, provides an exception to its general standards so that police and firefighters with 15 years or more of service are not subject to actuarial reductions for early retirement.

The February 17 edition of Roll Call reported that the Capitol Police Retirement Act, which became law on October 10, 1990, mandated that as of October 31, 1992, any officer 55 years of age or older must retire after completing 20 years with the Capitol Police force. The retirement plan, it was hoped, would encourage younger officers to remain on the force for longer periods as older officers vacated key posts. This is really important to us in the cities.

It was also seen as an opportunity to open up senior positions on the force to allow the promotion of, among others, minority officers.

Federal law enforcement officers must retire at age 57, and Federal firefighters must retire at age 58.

Public safety employment issues raised in 1986 remain important and have grown more important as violent crime has continued its upward trend, while many cities and towns have been forced to reduce the size of their law enforcement agencies. Simply quoting from the 1986 NLC statement, there are many reasons why State and local employees engaged in public safety activities should be exempt from ADEA, including the physically stressful character of the work and the difficulties involved in developing reliable, safe and cost-efficient tests of an individual's physical fitness. In the absence of clear standards governing the circumstances in which a mandatory retirement age of less than 70 years of age can be used for police and firefighters, the ADEA will hamper the provision of basic public safety and fire protection services.

The 1986 legislation was crafted so that this exemption would expire after the Federal Government, through the Department of Labor and EEOC, had conducted extensive studies of the subject with broad-based consultation with the groups affected. In fact, it was one of the compelling reasons for the 1986 public safety exemption from the ADEA.

During the seven-year exemption the EEOC was mandated by Congress to conduct a study to address the following issues, and I think that you have already addressed those here today. I am sure you have.

In September 1990, the EEOC awarded the study's contract to Pennsylvania State University [PSU]. As you know, PSU only partially completed the mandated study. Nevertheless, PSU presented a report to EEOC in December 1991 with the recommendations that public safety positions no longer be exempt from ADEA. It is important to point out that the PSU study and report failed to satisfy the most important parts of the congressional mandate. Rather, it only fulfills part (1) without addressing the more difficult areas of the study and the real-world business necessities mandated in parts (2) and (3)—specifically, to identify the particular types of tests to measure mental fitness and to develop specific standards under which to administer these tests, part (3). Consequently, unless permanent exemption from ADEA is implemented, public safety agencies will be left in a highly compromised, no-win situation almost identical to where we were prior to 1986.

The time schedule set forth in the statute provided tight, but realistic time frames in which to conduct the studies; and if testing procedures were determined to be better than fixed entry or retirement ages, the statute requires that specific tests be identified and guidelines for their use be developed.

This has been determined not to be the case. The last of these projects was to be completed more than two years before the expiration of this exemption. Again, as noted above, this did not occur. I am sure when the time schedules were laid out, they were created by Congress to allow Congress ample time to answer and deliberate on the results of these studies and reports.

The fact that these studies, reports and guidelines have not been completed destroys the sequence of events that was carefully incorporated into the statute. It is unfortunate that the expiration of the exemption was not made contingent upon the results and satisfactory completion of these efforts; but it was not, and therefore legislative action is required now.

I am sure that Congress never considered the possibility that its directives and accompanying time schedules so specifically provided for in statute would not be complied with. Allowing the exemption to expire without any reasonable prospect of having recommended tests, or series of tests, validated, once again opens up public employers to lawsuits contesting the effects of fixed maximum entry or retirement ages. Fostering needless litigation has placed towns and cities into a position where scarce resources will be used on legal costs.

In testimony—I am trying to condense this.

I think that I am going to go to the end of my testimony. I think that you have copies of my written statement. Good.

I am going to thank you for your attention, very much. I apologize for being late. As I said earlier, I was otherwise occupied earlier, and I know that the League of Cities will look forward to working with you on this issue and others in the future. Thank you for your attention.

[The prepared statement of Ms. Ticer follows:]

Honorable Patricia Ticer

Good Morning, my name is Patricia Ticer, Mayor of Alexandria, Virginia. I am here this morning on behalf of the 150,000 municipal elected officials and municipal organizations in 49 states represented by the National League of Cities.

The National League of Cities appreciates the Subcommittee's holding of this timely hearing to consider a subject of importance to cities and towns across the country and to the law enforcement officers and firefighters who provide essential public safety services to these communities.

Our purpose in presenting testimony is to urge the subcommittee to approve legislation that would allow ALL cities and towns the option to use age-based criteria for hiring and retirement of public safety personnel. Further, we would urge members to include language that would relieve municipalities of liability resulting from the expiration of the ADEA exemption on December 31, 1993.

As the Subcommittee is aware, the seven year exemption of police and fire personnel from the Age Discrimination in Employment Act (ADEA), as it applies to hiring and discharge only, expired December 31, 1993. This means that cities and towns have been left unable to enforce age-based hiring and retiring decisions based.

Over the past several years, NLC has worked hard in partnership with national organizations representing rank and file law enforcement officers and firefighters - labor and management together - to ensure that ALL cities and towns have the right to determine for themselves the types of hiring and retirement policies that meet the needs of each community. Such policies are essential for municipal elected officials to fulfill their legal responsibility to provide for the public's safety.

Last year, HR 2722 was approved by the House of Representatives without controversy. The Senate refused to take up the measure. The effects of this failure have resulted in a serious disruption in long standing employment policies and practices for those cities included in the exemption. Those cities and towns not covered by the exemption - those that did not have age-based policies on the books prior to 1986 - have been forced to implement policies that do not represent the best interests of the residents of their respective communities.

Our request is consistent with current federal policies and practices with respect to federal public safety personnel. In 1974, the Congress applied the ADEA to the federal sector. However, Congress retained the mandatory retirement age of 55 for federal fire fighters and law enforcement personnel. NLC believes the exemption for federal public safety agencies from the ADEA requirements is fair. The same flexibility should apply to all levels of government.

Mr. Chairman, my remarks will focus on six specific issues that will explain why NLC supports legislation that would make permanent the exemption and expand it to all municipalities.

First, and perhaps most importantly, the bulk of city governments and units of organized labor that represent the law enforcement and firefighting personnel working for these communities support legislation that would extend to all cities and towns the option to implement age-based criteria.

Second, federal law provides other special recognitions of retirement ages for law enforcement and firefighting personnel.

Third, public safety employment concerns raised during the enactment of the limited exemption in 1986 remain important.

Fourth, the reports, studies, and guidelines that were to have been done by the Department of Labor and EEOC well in advance of the exemption's expiration in 1993 have not been completed.

Fifth, the absence of well developed, validated and tested alternatives is resulting in costly and needless litigation.

Sixth, the impact on affirmative action goals which are critical to serving our ever more culturally diverse communities is of great concern to municipal leaders.

I. On the issue of extending the previous exemption to include all cities and towns, there is an agreement between city governments and major segments of organized labor representing law enforcement and firefighting personnel which these city and state governments employ.

As you are well aware, this level of consensus does not always exist. NLC would urge the subcommittee to take advantage of a win-win situation by advancing legislation to exempt cities and towns from the ADEA.

II. Federal law in other places provides special recognition of the particular structure of public safety occupations so that an exemption would not be inconsistent with other statements of federal policy.

For example, section 415 of the Internal Revenue Code which governs pension issues provides an exception to its general standards so that police and firefighters with 15 years or more of service are not subject to actuarial reductions for early retirement.

The February 17, 1992 edition of Roll Call reported that the Capitol Police Retirement Act, which became law on October 10, 1990, "mandated that as of October 31, 1992 any officer 55 years or older must retire after completing 20 years with the Capitol Police force. The retirement plan, it was hoped, would encourage younger officers to remain on the force for longer periods as older officers vacated key posts. It was also seen as an opportunity to open up senior positions on the force to allow the promotion of, among others, minority officers." Federal law enforcement officers must retire at age 57 and federal firefighters must retire at age 55.

III. The public safety employment issues raised in 1986 remain important and have grown more important as violent crime has continued its upward trend, while many cities and towns have been forced to reduce the size of their law enforcement agencies. Simply quoting from the 1986 NLC statement,

> There are many reasons why state and local employees engaged in public safety activities should be exempt from ADEA including the physically stressful character of the work and the difficulties involved in developing reliable, safe, and cost-efficient tests of an individual's physical fitness. In the absence of clear standards governing the circumstances in which a mandatory retirement age of less than 70 years of age can be used for police and firefighters, the ADEA will hamper the provision of basic public safety and fire protection services.

IV. The 1986 legislation was crafted so that this exemption would expire <u>after</u> the federal government, through the Department of Labor and Equal Employment Opportunity Commission, had conducted extensive studies of the subject with broad-based consultation with the groups affected. In fact, it was one of the compelling reasons for the 1986 public safety exemption from the ADEA.

During this seven year exemption, the EEOC was mandated by Congress to conduct a study to address the following three issues:

1. To determine whether physical and mental fitness tests are valid measurements of the ability and competency of police officers and firefighters to perform the requirements of their jobs;

2. If such tests were found to be valid measurements of such ability and competency, to determine which particular types of tests most effectively measure such ability and competency; and

3. To develop recommendations with respect to specific standards that such tests and the administration of such tests should satisfy.

In September 1990, the EEOC awarded the study's contract to Pennsylvania State University (PSU). PSU only partially completed the mandated study. Nevertheless, PSU presented a report to the EEOC in December 1991 with the recommendations that public safety positions no longer be exempt from ADEA. It is important to point out that the PSU study and report failed to satisfy the most important parts of the congressional mandate. Rather, it only fulfills part (1) without addressing the more difficult areas of the study add the real world business necessities mandated in parts (2) and (3) - specifically, to identify the particular types of tests to measure physical and mental fitness, part (2), and to develop specific standards under which to administer these tests, part (3). Consequently, unless a permanent exemption from ADEA is implemented, public safety agencies will be left in a highly compromised, "no-win" situation, almost identical to where we were prior to 1986.

The time schedules set forth in the statute provided tight, but realistic time frames in which to conduct the studies, and, if testing procedures were determined to be better than fixed entry or retirement ages, the statute required that specific tests be identified and guidelines for their use be developed. This has been determined not to be the case.

The last of these projects was to be completed more than two years before the expiration of this exemption. Again, as noted above, this did not occur.

I am sure that when the time schedules were laid out, they were created by Congress to allow Congress ample time to consider and deliberate on the results of these studies and reports.

The fact that these studies, reports, and guidelines have not been completed destroys the sequence of events that was carefully incorporated into the statute.

It is unfortunate that the expiration of the exemption was not made contingent upon the results and satisfactory completion of these efforts. But it was not, and therefore, legislative action is required now. I am sure that Congress never considered the possibility that its directives and the accompanying time schedules so specifically provided for in statute would not be complied with.

V. Allowing the exemption to expire without any reasonable prospect of having recommended tests or series of tests validated once again opens-up public employers to lawsuits contesting the effects of fixed maximum entry or retirement ages.

Fostering needless litigation has placed hard-pressed cities and towns in a position whereby already scarce resources will be used on legal costs.

In testimony submitted by NLC before the House Education and Labor Subcommittee on Employment Opportunities in 1986, we attached three and one-half pages listing cases which had been filed under ADEA related to police and fire retirement ages. This same testimony provided a listing of divided court decisions on whether age could be considered a Bona Fide Occupational Qualification (BFOQ) under ADEA.

VI. The achievement of affirmative action objectives may be hampered for cities just as with the Capitol Police. Continued movement toward more highly qualified law enforcement and firefighting forces that "look like the communities in which they work" is a goal for many communities. As our communities grow ever more culturally diverse, we must reach out to include personnel that can carry out real programs of community policing. These goals, however, must be achieved under tight fiscal constraints and public safety forces that in many communities cannot grow in overall size because of these fiscal restraints.

Thank you for your attention, and I assure you that the National League of Cities would like to continue working with you to deal with this important issue.

Chairman FAWELL. Well, thank you. I did have the opportunity to read all of the statements last evening.

Ms. TICER. Good.

Chairman FAWELL. And you sometimes draw the conclusion that you can't get there from here. It seems—we obviously can, but it has been a tortuous path to try to come to a solution.

I wonder, Captain Wall, you are a bit on the defense, as I have indicated before. Your testimony was very interesting when we do have, as we have in this instance, management and labor being so close and unanimous, et cetera. Why is it, from your viewpoint at least, that we have apparently such overwhelming numbers of employees who are firefighters and/or law enforcement officers who seem to disagree with your view and are seeking age-based criteria for hiring or discharge retirement?

Captain WALL. I cannot attest to any polls that have been conducted—or I have never seen any—that states there is an overwhelming majority, other than what I read in the local international paper that I get. But I can tell you that the department I work on and I work with—there are over 2,400 men on our job, and the majority of the people I work with do not agree with mandatory retirement.

Chairman FAWELL. Have you had a poll at all?

Captain WALL. No, I haven't conducted a poll.

Chairman FAWELL. And I assume that others have not. Or has there been a poll? It would be interesting, if indeed there has been a poll of membership, for instance.

Mr. Nesbitt?

Mr. NESBITT. We did not conduct a poll, but I can tell you that we can represent the local and—the Los Angeles Local 1014, and in that case, the membership has voted in support of this position. The union leadership supports it; the fire department and the fire chief of this department also support the public safety exemption from ADEA.

Mr. SCULLY. Also from the police perspective, we have annual conventions at which all legislative resolutions are discussed. We just don't create these resolutions and come in here and pursue legislation unless we have a direct mandate from the elected leadership—in my case, of nearly 3,500 police associations throughout the country. And of the 37 legislative resolutions that were passed at my convention in August, I can tell you that this resolution was passed unanimously; and they are speaking for the members that elect them on a local, county, or State level.

Ms. TICER. It is the same, of course. Our policies are a matter of discussion in committees and then full vote at our national meetings. So, it is representative of the local governments.

Chairman FAWELL. I also note that Mr. Nesbitt in his testimony and in his written statement had stated that only one-seventh of all firefighters are over the age of 50, but that they account for more than one-third of all firefighter deaths.

Assuming those figures are accurate, that is relatively impressive, at least to this one Member of Congress. Being one who has passed the age of 60, I tend—just out of the normal course of my life—to believe that other than some very unique individuals such as yourself, that age-based criteria, when you consider the tortuous paths that employers have to go through to find a fitness test that will pass judicial scrutiny, may be the most fair course that we could take.

How would you respond to that?

Captain WALL. Well, I don't know. I have not—I have read about firefighters being killed in the line of duty. But I don't know how many actually are over the age of 50, if it is as great a number as the union is saying or not. I can tell you about the South Canyon fire in Colorado out at Glenwood Springs this past summer, one of the largest fire disasters in our Nation's history, where 14 wildland firefighters burned up.

There was, I believe, two people over 30 years of age on those crews. And I am not saying that if they had had somebody with more experience, more years on the job if they would have been in that same position or not. But I don't really think there are any hard, fast conclusions that an individual over 50 years of age is more susceptible to being killed on the job than somebody younger.

Chairman FAWELL. All right. Thank you.

Captain WALL. I would also like to address Mr. Nesbitt's last statement about the Los Angeles County Fire Local. Union 1014 took a poll. I am a member of Local 1014 and I never received any information on voting on mandatory retirement through our local. And I know our fire chief—he is right, our fire chief—there is a statement come out from our department that they support mandatory retirement. Whether they will reenact it if this legislation is passed, I don't know.

Mr. NESBITT. I didn't say we took a poll. I said the membership voted and the leadership supports mandatory retirement.

Chairman FAWELL. On the legislation you are referring to?

Mr. NESBITT. Right. And also at four elections, where our members are democratically elected, all four times they have supported this position on mandatory retirement.

Chairman FAWELL. I thank you. I see my five-minute period has expired.

The Chair recognizes the gentleman from California, the Ranking Minority Member.

Mr. MARTINEZ. Thank you, Mr. Chairman.

Fred, just to follow up on that last question a little bit. When you say Local 1014 voted, they voted as a local in the union in the hall?

Mr. NESBITT. I assume it was at a regular meeting.

Mr. MARTINEZ. There must have been some result, like so many for and so many against.

Mr. NESBITT. That I don't know, but I can get it for you.

Mr. MARTINEZ. It would be interesting to see how many voted against it.

See, I have—and you might expound on this, Fred. You mentioned in your oral statement that this bill gives the flexibility to local agencies to make that determination, local city councils to set a policy or develop a law. Well, years ago, before the Age Discrimination in Employment Act was passed, we in Monterey Park, when I was the Mayor, did have mandatory retirement of 55 because it was the standard policy at that time. I think most departments did, but there are probably exceptions. And yet when a person like, let's say, Barney became of that age, we didn't retire him because he was an exceptional individual.

You know, in your testimony, Captain Wall, you said you are not unique. You are unique. There are not many people at 60, 61, can do the things that you do. And you mentioned the other firefighter, the captain who is 67 or 65; that still is unique. I mean, you are a couple of people, and I imagine if you took a count of those kinds of people, there are going to be very few.

Now, I am one of the opinion that if you have an individual like that, you ought to keep them on mostly because of what you said about his experience and the ability to maintain calm in a crisis situation.

I remember years ago when I was just a young Marine myself, and I was only 19 years old and most of the men in that unit that I was with were the same age, and we were called out for a firefight with the Chinese Communists, and none of us had been in action. The older ones, Master Sergeant Stein and Mason, they had been in combat through World War II, and they knew what to do. If it hadn't been for them maintaining the calm—we were scared out of our minds. We didn't know what we were getting into. They did maintain calm, so there is something to be said for that experience.

But in this law we give the flexibility to the cities, and I would ask Ms. Ticer if in your situation you found an individual like Barney.

Let me tell you about Barney. Sergeant Burke, who is probably 20 years to his junior, challenged him to a 100-yard dash and guess who won. Barney did. And then he challenged him to a mile run and guess who won? Barney did. And with that, the Chief told Sergeant Burke, you better get in shape, Buddy; and that is true, Sergeant Burke had let himself go. And age—I have said this before, earlier on, you can let yourself go and not be in shape.

You mentioned in your testimony, Captain Wall, that you never started to do all of these exercises until the department started mandating this.

Ms. TICER. We would probably—we probably would like to have the flexibility to make the exception. You know, it is an exception though. It is an exception. And I think that the real talent lies in being able to know, when can you do it and when you cannot.

And I would suspect that three-quarters of the men and women would know when it was time to retire voluntarily, and maybe a quarter of them would not. And so some of the people who should not be staying on might stay on, and then there would be some difficulty.

Localities have a lot of problems. They don't have all of the flexibility of businesses in hiring and firing. It is very difficult to do.

It is very important from a public safety—we have a very active volunteer fire department which subscribes—that works with the actives, and this I have promoted, again because they were out of favor for a while. And those people can provide all of the encouragement and the sageness that you are talking about on the side, without actually having to lift the heavy ladders or the heavy hoses or whatever it is.

But I would remind you of the difficulty in the standards. That is what we are talking about and that is what was supposed to have happened. But I think as important for us is the ability to have upward mobility within any department of government. You lose good people if there is no opportunity for upward mobility. And if there is no rotation on and off, then you don't have that. You don't have for your goals of affirmative action—you know, any community has multiple goals—public safety, of course, being the highest goal that we have.

So, I would say that we would like to have the flexibility, but as I said in my testimony, there must have been some reason and study behind the Federal standards, the Federal retirement at 55 and 57.

Mr. MARTINEZ. Actually, Mayor Ticer, there was not.

Ms. TICER. There wasn't? I am sorry; that was an assumption. I say that on the side; I shouldn't say it frontally.

Mr. MARTINEZ. I am wondering, Mr. Chairman, are you sure you didn't give me a short block? The time sure goes fast.

Let me say that Barney did retire at the age of 67 last year.

Chairman FAWELL. I thank the gentleman.

The Chair will recognize a new Member of Congress, Dr. Weldon, a physician, and I look with great anticipation to hear what this gentleman has to say with regard to the topic under discussion.

Dr. Weldon.

Mr. WELDON. Thank you, Mr. Chairman. Being that I am the only one here, does this mean that I get 35 minutes?

Chairman FAWELL. We will take that under consideration.

Mr. WELDON. Well, I would like to add for the record that while there are some seniors who, through very vigorous physical training, are able to maintain their physical performance, there is a physiologic decline in physical performance with age that is really very well documented.

Cardiac output declines with advancing age, pulmonary function testing declines with advancing age, and indeed, reaction times also decline with advancing age. Now, there are some individuals who can overcome that age-related decline, and there are some individuals, indeed, who don't seem to experience that age-related decline. They seem to have a genetic predisposition to maintaining physical fitness. Many of those individuals live to very advanced ages. So there is a tremendous amount of variation in the population.

But for the record, the use of age criteria in the past has been based on age-old common sense, based on observation of the natural physiologic phenomena.

And indeed, Mr. Wall, maybe you can comment on this. The purposes of these laws or regulations are not, per se, to protect or defend individuals, but to promote the public good in general.

Would you agree with that, while these types of regulations may adversely affect individuals such as yourself, there may be a greater public interest involved in having age standards?

Captain WALL. Definitely there is a public interest that is involved. And I read in—I believe it was the Penn State University study, I read a portion of that study that stated that the aging process starts at various times in an individual's life. Some people in their 40s are aging faster than people up in their 60s and 70s.

And I would just say that I would like to see everybody be able to provide the service to the public in a manner, you know, that they are capable of doing. If you are 40-some-years old and you can't do it, you shouldn't be there. If you are 60, you shouldn't be on the job. And that can all be done through testing, I feel.

Mr. WELDON. Mayor Ticer, has the City of Alexandria used age-based criteria in the past?

Ms. TICER. Yes, it has.

Mr. WELDON. And has the city been involved in any litigation?

Ms. TICER. The city, yes, but I am sorry that I didn't do my homework well enough on this subject this morning, so I can't give you the specifics on that. But in a couple of instances, since the law lapsed, we have had at least grievances. And I am not sure whether it went to litigation or not. I am sorry that I cannot answer that. I can get you that information, though.

Mr. WELDON. I would appreciate it if you could.

Ms. TICER. I will.

Mr. WELDON. I have another question for Mr. Wall. Most States have criteria—age criteria for obtaining a driver's license for a youth. And would you agree that those laws tend to be somewhat arbitrary and that there would be exceptions where some people perhaps who would be under the age limit would perhaps have the maturity and the skill to be able to obtain a driver's license but that it would perhaps not be in the public interest to allow people of any age to get a driver's license?

Captain WALL. I can relate to that, yes. I was in that position when I was at that age.

Mr. WELDON. That is all of my questioning. Thank you.

Chairman FAWELL. Thank you. I thank the gentleman for some very fine comments.

The Chair recognizes now the gentleman from New York, Major Owens, who I think probably of all the Members on this particular subcommittee is as conversant and deeply aware of all the nuances and the ins and outs of this particular issue.

Mr. OWENS. Mr. Chairman. Thank you.

I have a written statement that I would like to submit for the record.

Chairman FAWELL. Without objection.

[The prepared statement of Mr. Owens follows:]

OPENING STATEMENT

Congressman Major R. Owens

Hearing on the Public Safety Exemption from ADEA

January 24, 1995

It is imperative that the Congress move swiftly to enact legislation to permanently exempt state and local public safety agencies from the Age Discrimination in Employment Act in order to permit them to consider age in their hiring and retirement policies. Such legislation passed the House unanimously on two occasions last year but was prevented from even being considered by the Senate by the threat of a filibuster. There must be no further delay. Let us get on with it.

As a rule, Congress must avoid exempting whole classes of employees from the protection of civil rights laws unless it is absolutely necessary. We should not carve out exemptions merely because an employer finds civil rights compliance to be costly or inconvenient. Exemptions must be made only when there is a strong compelling need to do so and there is no other reasonable alternative. This is one of those rare instances.

State and local fire and police agencies must be exempted from the ADEA in order to protect and promote the safety of the public. This is literally a life or death matter. If a police officer or firefighter cannot adequately perform their duties, people die and people get hurt.

Age does indeed affect an individual's ability to perform the duties of a public safety officer. This is not a stereotype. This is not ageism. This is a medical fact. Physical ability declines with age. For example, aerobic capacity declines at a rate of 1% per year after age 30. Strength declines at a rate of 10-13% every decade. The risk of sudden incapacitation also clearly increases with age, increasing sixfold between the age of 40 and 60 years of age. These physical effects are not experienced by all people to the same degree or at the same precise time. But they pose a significant problem to public safety agencies in their efforts to maintain a fit and effective workforce.

A public safety agency can respond to age-related declines in ability in one of two ways. It can establish an age-based mandatory retirement policy. This will reduce the risks to public safety, but it may result in some capable individuals being forcibly retired.

Alternatively, an agency can try to use performance and physical ability testing to try to screen out employees who might pose a threat to public safety. Unfortunately, there are numerous problems with trying to use tests as an alternative to age which makes this option untenable.

It is simply not possible to devise a test for all tasks carried out by a public safety employee. For example, no test could have possibly simulated the kinds of physical conditions public safety employees in California have faced over the past few weeks of severe flooding. No test, no matter how comprehensive, can measure all of the skills and abilities a public safety employee must possess.

Moreover, there is no current test that can effectively screen for the risk of sudden incapacitation among asymptomatic individuals. A mandatory retirement age, used in conjunction with screening for other risk factors, continues to be the most effective way of reducing the risk of sudden incapacitation by public safety officers.

Testing can also have a very serious negative impact on other individuals and groups that historically have been discriminated against in employment. Tests have been proven to have an adverse impact on women and minorities. Women on average are less strong than men. Written tests may underpredict the on-the-job performance of minorities. To assure that such factors did not prevent women and minorities from serving in public safety positions, many agencies "within-group" normed the results of certain tests. Unfortunately, a provision of the Civil Rights Act of 1991 now prohibits that practice. As a result, any increase in the use of physical and mental testing of public safety employees will jeopardize employment opportunities for women and minorities.

Another, but lesser concern is that it is enormously expensive to administer performance and ability tests on a periodic basis to all public safety employees, consuming scarce resources that are needed to keep police on the streets. In addition, testing often entails considerable litigation over the content of the tests. In Tennessee, for example, there were several years of litigation over the state wildlife officer's entrance exam which focused on the question of whether the fences recruits had to scale should be 8 or 10 feet tall.

For these reasons, testing does not today represent a viable alternative to age-based mandatory retirement policies for public safety agencies. If public safety agencies are exempted from the ADEA, those agencies who wish to experiment with testing in lieu of retirement ages will be able to do so. But given the uncertainty about the effectiveness, effects and implications of using tests as a substitute for age, the Congress must not force every public safety agency to implement them. This would be the effect if we did not enact an exemption.

I urge the members of the Subcommittee to join me in supporting the enactment of a public safety exemption to ADEA. It is imperative that all public safety employees be fit, effective and fully capable of fulfilling their duties. An ADEA exemption will assure that state and local police and fire agencies will be able to pursue that goal using the same age-based employment criteria which is now used by the FBI, the Secret Service and other Federal public safety agencies.

Mr. OWENS. As you I stated, as a sponsor of the legislation in the past Congress, I am quite familiar with the various aspects of this issue; and we did give to all Members, I think, a packet which contains the legislative history and some other facts which I think would help the new Members of the committee in making a decision on this matter.

Nobody supports more strongly any effort to end discriminatory rules and regulations and practices than I do. And in this case, I think certainly age discrimination is a serious matter. There are exceptions to every rule, and in this case, exceptions to a principle; and the need for this exception, I think has been very well documented.

The doctor's statements before, of course, are consistent with what most other physicians are saying also in terms of there—it is just a medical fact that age has an impact on one's ability to function in this kind of a position where your physical fitness, your physical strength, all of that is of major importance. Then there are other factors introduced into firefighting and law enforcement related to stress and emotional strain which increase the necessity of fitness.

Common sense all around us tells us that there is a definite point at which people just cannot perform, even the best, as they age. They cannot perform the same as they grow older as they did when they were younger. Athletes don't dispute this. Most athletes in even the sports that are not contact sports recognize that there is a point where they want to get out of competition. And they are usually looking for a position in the front office if they want to stay close to the sport. Certainly, baseball and basketball, big issues have been made that we want to continue, but we want positions in the front office.

So I want to raise a question with Captain Wall. To what degree did your position as Captain involve administrative and, you know, management work, where you were really not performing firefighter duties most of the time?

Captain WALL. Well, little of my time—a lot of my time is administrative, but the majority of it is actually working out in the field with the crews and on the emergency incidents.

Mr. OWENS. How many captains were there in the department? You are retired now, right?

Captain WALL. No, I am still working.

Mr. OWENS. How many captains are there?

Captain WALL. There are over 600—640, I believe, is the last figure.

Mr. OWENS. In a force of how many firefighters?

Captain WALL. Two thousand four hundred total uniformed personnel in our department.

Mr. OWENS. Out of 2,400—the total personnel is 2,400 and 600 are captains. So a large number of them can aspire to become captains and have some of their workload entail administrative and managerial work, but most of them cannot. If they stayed on without any kind of mandatory retirement, they cannot become captains because there is not enough room in the captain ranks for all of those people. Am I correct?

Captain WALL. There—you mean the opportunity for promotion is not there?

Mr. OWENS. I mean, are we dealing with an issue of, given the fact that you could become the captain, you are not performing the same duties as you performed in the department when you came on the force, so there is already a factor here which means that you may be able to go out and periodically work at the same level as you worked before, but you are not called upon to do that with the same frequency as you were when you were an entry level firefighter?

Captain WALL. Yes, I do. I work at a fire suppression camp. All wildland fires in the County of Los Angeles, I have a good chance of responding to it right now with the crews—putting on my pack, all my equipment, climbing up the steep rugged terrain.

Mr. OWENS. While you are working with that emergency. But in routine ways, a large part of your work is administrative?

Captain WALL. Yes, a large part of it is.

Mr. OWENS. The assumption is that as you grow older and are promoted, certainly you can make that adaptation. Of course, you as an exceptional individual do not have to make that adaptation, but in most cases people who are growing older want different types of jobs. And you have a managerial problem that is on top of the problem that we are discussing here, that most firefighters are expected to stay in the same capacity as they were in when they came in and enduring the physical impact of that work.

So we cannot assume that they can make that kind of adjustment; and I think it is important to make that point, that there is just no room for large numbers of people to change and even to some degree downgrade the amount of exertion, stress that exists on the job.

Captain WALL. In our department, we have such a variety of different types of jobs. We have wildland areas; we have residential, commercial, industrial work to do; there is a Fire Prevention Bureau where a lot of the younger captains who want to bid into that position, they do it on a seniority basis, and there are a lot of people that transfer into a position where they feel comfortable and want to work. If they feel like they don't want to work in the field any longer and do the emergency work in the field and at the station, they go to a position in the Fire Prevention Bureau, for example.

Mr. OWENS. That is the point I wanted to make, that whole assumption that they can take jobs which are not as stressful and not the same as the regular firefighting jobs or regular law enforcement jobs, that is what we kept running into repeatedly. That is a different kind of problem, and make the systems need to make room for them to step into jobs that are primarily managerial and no longer have the same kind of stress. That is a different kind of problem.

But basically the people that are out there in the position of doing the express work that they were hired to do cannot continue to do that after a certain age.

Captain WALL. That is right. I could bid into one of these spots if I wanted to now. I am number eight on the seniority list in our department out of 2,400, but I don't want to. I want to work in the field in fire suppression until I can't do it any longer because I don't like office work and paperwork and administrative duties like that as well as I do working in the field. And I just feel, you know, if we are serving the public in an outstanding manner, age shouldn't have anything to do with our abilities to perform those duties.

Chairman FAWELL. Well, I thank the gentleman for his questions and comments. And I thank all of the participants. Your testimony has been excellent. I know it has helped me a great deal certainly to refresh my recollection, as we lawyers would say. And if ever there is proof that the law—at its best—is imperfect, we have this particular situation; and all we can do is our very best to try to solve the situation.

We will make another effort to see if the Chairman of the EEOC can get into that study and, hopefully, have a better ending than has been the case with that initial study that was engendered back in 1986. Maybe we will find that there are fitness tests that really are able to determine physical and/or mental ability. And I gather that we are all aware that the legislation does not necessarily mean, as has been indicated, that a person of your talents cannot be worked into the local law. Although I think we all understand when you start making an exception for a man of your unique talents, sometimes then we are back in the thicket of discrimination cases and things of that sort because there may be those who feel that they were not given the recognition that you were given.

So, thank you very much for your testimony. We do appreciate it.

And the meeting is adjourned.

[Whereupon, at 10:54 a.m., the subcommittee was adjourned.]

[Additional material submitted for the record follows.]

INTERNATIONAL BROTHERHOOD OF POLICE OFFICERS
AFFILIATED WITH THE SERVICE EMPLOYEES INTERNATIONAL UNION, AFL/CIO
2011 Crystal Drive, Suite #206
Arlington, VA 22202
Telephone 703/979-0290
Fax 703/979-0294

Testimony of the

International Brotherhood of Police Officers

to the

Committee on Economic and Educational Opportunities

Subcommittee on Employer-Employee Relations

regarding

The Public Safety Exemption

to the

Age Discrimination in Employment Act

January 24, 1995
2175 Rayburn

I. Introduction

The International Brotherhood of Police Officers (IBPO) is and affiliate of the Service Employees International Union, the fourth largest union in the AFL-CIO. The IBPO represents more than 40,000 federal, state, and local rank and file law enforcement officers across the country, and is the largest police union in the country. As a labor organization, the IBPO represents street cops--those who each day put their lives on the line to protect their communities. The IBPO through its membership strongly supports legislative efforts to permit the use of maximum hiring ages and mandatory retirement ages for state and local firefighters and law enforcement officers.

We wish to thank you, Chairman Fawell, for holding this hearing today on this important topic. Your support for this issue during the 103rd Congress is much appreciated, and we look forward to working with you to write the exemption into law this year. As you know, support for this issue is widespread. Legislation accomplishing our goals passed out of subcommittee and full committee in the House unanimously. The full House of Representatives considered this issue twice last year, as a free-standing piece of legislation on the suspension calendar, and as an amendment to the omnibus crime bill. Unfortunately, despite support from many Senators, the ADEA exemption was prevented from becoming law. The IBPO pledges its support to removing whatever obstacles remain to swift passage of an ADEA exemption.

Although an ADEA exemption affects both firefighters and police officers, we will confine our testimony to the impacts on police officers. The IBPO supports an amendment to the Age Discrimination in Employment Act which allows state and local governments to utilize mandatory retirement ages because: (1) age provides a more reliable indicator of physical ability than physical and mental fitness tests; (2) age based criteria protects public safety and the safety of other officers; (3) age provides a workable objective standard for retirement; and (4) a statutory exemption enables police departments to avoid costly litigation with disparate results.

II. Legislative Background

The Age Discrimination in Employment Act of 1967 (ADEA) (29 U.S.C. §§621 et seq.) legislates that employers must not use mandatory retirement ages to force employees to retire. However, because of a 1986 Congressional amendment, police and fire departments were allowed to continue to use mandatory retirement ages. Also in the 1986 amendment, Congress mandated a study to determine the efficacy of fitness tests, which we will discuss in our testimony below. The 1986 amendment included a seven year exemption from the ADEA which expired on December 31, 1993. This exemption should be reinstated and applied retroactively so that state and local governments may continue to use age-based criteria for retirement decisions.

One reason behind Congress' decision to carve out this exemption from the ADEA in 1986 is that it would be unreasonable to require state and local governments to be held to more restrictive laws than the federal government. Congress has specifically mandated a retirement age for federal firefighters, U.S. Park Police, the Capitol Police, and many other categories of federal public safety officers. Mandatory retirement ages have worked well in the federal law enforcement sector. We believe state and local governments should likewise be allowed to establish age-based criteria for their public safety officers.

III. Why an Amendment to the ADEA which would allow use of age-based criteria is necessary

a. Age currently provides a more reliable indicator of physical ability than physical and mental fitness tests

In 1986, Congress mandated the EEOC and the Department of Labor to conduct a study to determine if physical and mental fitness tests are valid measures of ability of police and firefighters to perform requirements of their jobs, which tests are most effective, and which specific standards such tests satisfy. Researchers from Penn State conducted this study, which concluded that although age is a determinant of an individual's ability to perform the functions of a police officer or a fire fighter, there is no exact age at which all people can be said to be unqualified. While we agree with this statement, it does not necessarily follow that age-based retirement ages are illegitimate.

Tremendous practical problems exist with the summary conclusions of the Penn State Study. For example, while the study recommended the use of fitness tests as an alternative to age-based retirement, it failed to identify any specific physical or mental fitness tests. Further, it failed to even offer any recommendations as to specific standards that should be satisfied by such tests. The EEOC also failed to promulgate guidelines to assist state and local governments to administer and use physical and mental fitness tests instead of age-based policies.

The IBPO, therefore, has several questions which must be answered conclusively before Congress mandates state and local governments to use tests. Can a test simulate the entire range of public safety tasks including an individual's response to mass civil disobedience and other emergency situations, or to combative, resistive arrestees? Can such tests effectively screen out individuals who are at risk of sudden incapacitation? The Penn State study completely failed to answer these questions. The IBPO believes that currently these questions do not have legitimate answers.

Not only did this study fail to determine if mental and physical tests could be effectively used, it also was based on significant errors. A comprehensive review of the study found serious flaws in data collection, data analysis, and study selection. Most significantly, the study's recommendations are based on inferences that are not supported by any data. In real world situations, as opposed to laboratory settings, age has proven to be a more reliable indicator of physical fitness than medical testing.

Though the Penn State Report looked at performance tests used by law enforcement agencies to select entry level personnel, it did not evaluate the effectiveness, reliability or validity of such tests generally. Incredibly, the study failed to evaluate the appropriateness of the use of such tests to evaluate incumbent public safety officers. The Report itself even notes that many of the research studies on efficacy of testing must be "interpreted with caution" because such studies did not include older individuals in the sample group. The Report goes on to say that most of the studies tested only individuals under age fifty. The IBPO finds it absolutely incomprehensible that a study can claim to make conclusions about what older people can or cannot do when the researchers have not even studied older people!

b. Fiscal Costs of Implementation if Tests are Imposed

The costs of these as of yet unidentified tests would include not only the administrative costs associated with developing and administering tests, but also the costs associated with a defense of the testing program and standards in the context of litigation. Even the Penn State Report found that it is likely that legal challenges to retirement decisions will increase if tests replace age as the criterion. We might note that numerous lawsuits have been filed within the last year when the exemption expired. We must restore the exemption quickly or these costs will continue to escalate.

c. The Use of Age-Based Criteria Protects Public Safety and the Safety of Other Officers

There are many compelling reasons for continuing to use mandatory retirement ages for police officers. While the IBPO supports the intent of the ADEA, we believe that special consideration must be given to safeguarding public safety.. Because officers perform tasks requiring great physical exertion and extreme stress, they must be in reliable physical shape. There are serious consequences when police officers fail to properly respond and control critical incidents, including serious physical injury or death of citizens, and significant property damage. Police departments must always be prepared for the worst case scenarios and have sufficient numbers of individuals qualified and able to meet the challenges, no matter what the crisis.

Older public safety officers are more likely to suffer catastrophic medical events, such as heart attacks, thereby jeopardizing the safety of their fellow officers, as well as increasing more expensive duty disability retirements. Because older people slowly lose essential, job-related skills over time, the safety of fellow officers and the public-at-large may be placed at risk. It is hardly realistic to expect 65 year old officers to expend the amount of physical energy necessary to subdue young and violent criminals. Even the Penn State study found that if state and local public safety agencies did not use age-based criteria, an estimated 8.75 events per 100,000 public safety officers per year would occur in which people die or are injured as a result of the sudden incapacitation of a public safety officer.

If police officers cannot adequately perform their duties, people could die or get hurt. The House Report to accompany H.R. 2722 reported: "Age affects an individual's ability to perform public safety duties: that conclusion is neither a stereotype not an unproven assumption; it is medical fact." (H.Rpt. 103-314, at 9.) This medical fact is derived from the very same report that vaguely concluded that age is an imprecise predictor of decline in ones ability, the Penn State report. Indeed, despite its conclusions, the Penn State report found: physical ability declines with age, e.g., aerobic capacity declines at a rate of 1 percent per year after age 30; strength declines at a rate of 10 to 13 percent every decade; and, the risk of sudden incapacitation increases six-fold between the age of 40 and 60.

We recognize that Congress should act with caution when exempting categories of employees from protections of civil rights laws. However, if older police officers who are unable to perform their jobs are retained on the force as required by the ADEA without this amendment, people will die or get seriously injured. The lives and property of citizens must not be sacrificed to obtain an appearance of nondiscrimination. Protecting public safety is a compelling reason to justify this exemption.

d. Age Provides a Workable Objective Standard for Retirement

While we appreciate the position of those opposed to an exemption that tests prevent discrimination, the IBPO submits that more discrimination will result if this exception fully expires. A mandatory retirement age is preferable to subjecting officers to yearly physical fitness tests not only because of the above mentioned inherent problems with tests, but also because the use of these tests may be left to the discretion of department heads to prematurely force officers out of their jobs. Additionally, the potentially negative impact that tests may have on individuals and groups that historically have been discriminated against in employment is well documented. Tests can be used inappropriately to target or harass employees and are the source of great anxiety for many employees. Thus with all the problems with these tests, it is much preferable to have a set age limit at which all officers will know that once they reach that age, they will retire.

e. A statutory exemption to the ADEA enables police departments to avoid costly lawsuits with disparate results.

With all the evidence pointing to the use of age as a more reliable indicator of physical ability than fitness tests, state and local governments seeking to protect public safety will continue to rely on age based criteria, just as they did prior to the 1986 amendment to the ADEA which exempted police and fire

88-349 O - 95 - 3

departments. If the statutory exemption to the ADEA is not continued, departments that wish to use a mandatory retirement age will have to prove that age is a Bona Fide Occupational Qualification (BFOQ), reasonably necessary to the normal operation of business. Parties to a dispute over the use of retirement age limits have been forced to litigate their disputes in court. The results from cases litigated before the 1986 exemption have been wildly different and caused great uncertainty within police departments as some jurisdictions found limits permissible and other jurisdictions found the very same limits impermissible. In addition to causing legal disarray, lawsuits over BFOQs are extremely costly, causing both the employers and the employees to needlessly stretch their very tight budgets.

IV. Conclusion

Tests can not accurately gauge the ability of public safety officers to perform their duties. The Penn State study that concluded otherwise is seriously flawed - both in its analyses and the self interested motivation underlying its conclusion. Tests bring with them a great fiscal cost - in developing and administering, and defending against law suits. If police officers are unable to perform their duties, people will die. Public safety is too important to sacrifice just to obtain a thin veneer of protection of civil rights. For all these reasons, it is imperative that Congress enact this amendment and allow state and local police and fire departments to continue to use mandatory retirement ages.

Dr. Frank J. Landy
Professor Emeritus
Penn State University

INTRODUCTION

On January 24, 1995, the Subcommittee on Employer/Employee Relations of the House Committee on Economic and Educational Opportunities held a hearing to consider a public safety exemption from the Age Discrimination in Employment Act. The invited witnesses included representatives of three organizations supporting age-based mandatory retirement and one fire captain opposing age-based retirement.

I was conspicuously absent from the witness list. I use the term "conspicuous" because I directed the congressionally mandated study that resulted in the removal of the previous ADEA exemption for public safety officers. The results of that $1,000,000 study, commissioned by the Departments of Labor and the Equal Employment Opportunity Commission, were accepted by those agencies and forwarded to Congress as their official policy positions. I am certain that the file on this issue contains the September 22, 1993, letter forwarded to the Honorable William D. Ford from the Acting Chairman of the Equal Employment Opportunity Commission, Mr. Tony Gallegos, indicating the official position of that agency opposing age-based retirement for public safety officers. I have included a copy of that letter as <u>Exhibit A</u> of this testimony and ask that it be included as part of my submission for the record of these hearings.

As the Director of that study, a study on which Federal Agencies based policy statements, it is difficult for me to understand why I was not invited to speak at your recent hearing, particularly in light of the comments of the IAFF regarding the Penn State Study. My absence was also "conspicuous" in light of the fact that I had been invited to present study results and scientific background for the study in the previous hearing before both the House Subcommittee chaired by the Honorable Major R. Owens and the Senate Subcommittee chaired by the now-retired Honorable Howard M. Metzenbaum.

Nevertheless, I was not invited to present testimony at the Floor Hearing on January 24, 1995, but have been assured that my written comments will be included in the record. I would request that my full statement and enclosures be included as part of the published record and not simply added to the file of that hearing.

COMMENTS

In the hearing that occurred in the House of Representatives on March 24, 1993, chaired by the Honorable Major R. Owens, Frederick Nesbitt, a lobbyist for IAFF, presented a critique of the study conducted by Penn State on behalf of Congress. On April 2, 1993, I submitted a formal response to those criticisms. I append that response as Exhibit B. It is interesting to note that Mr. Nesbitt's testimony in the most recent hearing is virtually identical to his earlier testimony, with no apparent recognition of the earlier rebuttal. In the interests of time and space, I will not repeat those detailed rebuttals in this response but will refer the reader to Exhibit B which presents the responses to the IAFF "critique" in great detail and with specific citation.

Since I have received a copy of the written comments of Mr. Nesbitt at the most recent hearing, I will deal with the issues raised in that testimony, sequentially.

Nesbitt Testimony

P.1.: IAFF asserts that age is correlated with performance and that this relationship is confirmed by "...a close reading of the State (sic) study...". It is clear that Mr. Nesbitt did not conduct a "...close reading" of the Penn State Study, or if he did, he did not understand what he closely read. Our study states unequivocally that when life style, injury, and illness are separated from age, age is not substantially correlated with performance for public safety officers (or any other profession).

P.1: The "analysis" he refers to was not conducted by the House Subcommittee but rather by the IAFF. Our response to that "study" is contained in Exhibit B, attached to the current submission.

P.2: The IAFF "study" to which Mr. Nesbitt refers is a simple mortality study. It would surprise no one to learn that the most experienced fire fighters are those most at risk, simply because they place themselves in the most dangerous situations. But these "numbers" must be kept in context. In 1990, there were a total of 102 deaths, for all reasons, among fire fighters on duty. Of those deaths, 37 were due to heart attack, and only 16 occurred in fire fighters without previous symptoms of heart disease. Of those 37, fewer than 16 occurred on the fire ground. We know that fewer than 2% of current fire fighters are 60 or older. The simple arithmetic suggests that we would expect a total of less than 1 death per year on the fire ground for fire fighters 60 or older. Further, even this death might have been prevented with a good cardiac screening program, such as we have suggested in our report.

The point is that IAFF is asking Congress to engage in active age discrimination based on an anticipated death rate of less than 1 (preventable) death per year. How can this be reconciled with the violation of the constitutional rights of public safety officers over the age of 40 and the rights to property and personal protection of citizens deprived of the experience and ability of fire fighters forced to retire even though their job performance is exceptional?

P.2: Mr. Nesbitt claims that support for the legislation is broad and unanimous among "the workers." As he well knows, this is not true. In fact, various workers' organizations have presented statements opposing the legislation for the record. Among these groups are the Massachusetts State Police Commissioned Officers Association, the Massachusetts Police Association, and the Police Executive Forum. Their official positions are presented as Exhibits C, D, and E. In addition, opposition to the proposed legislation among those in IAFF leadership positions has been crushed when it has appeared. As an example, the IAFF State Area Representative in California and President of an IAFF local, Mr. Brook L. Beesley, voiced opposition to the legislation and the IAFF position on this legislation and was summarily removed from office. If you gag your internal opposition, "unanimous" support loses its meaning. I have attached the documentation regarding the Beesley case as Exhibit F.

Finally, I have received and continue to receive a stream of phone calls from fire fighters and police officers endorsing the results of the study and offering to take any fitness test necessary to demonstrate their continued competence. Captain Wall, who testified before the Subcommittee on January 24, 1995, is a dramatic example of the class of public safety officers over the age of 40 who work hard to maintain their fitness and continue to perform at extraordinary levels, levels higher than any of their younger counterparts.

P.2-4: These testimonials are no doubt sincere, but do not represent the Captain Walls who remain capable. No one has suggested that standards be abandoned -- only that fitness be determined by performance rather than by age.

P.4: As indicated above, there are organizations who oppose the legislation (documentation has been presented in Exhibits C, D, and E). Further, it is telling that Mr. Nesbitt can list no scientific body or organization that either supports the proposed legislation or criticizes the Penn State study and recommendation. The IAFF is a labor union not a research organization. The results of our study have been reviewed and endorsed by members of the appropriate scientific groups such as the American Psychological Association and the American College of Sports Medicine.

P.4-5: IAFF calls for a "balanced" workforce. We agree. A balanced workforce is balanced by age, experience, gender, ethnicity, etc. It is hard to understand how arbitrarily eliminating an effective segment of the work force (older public safety workers) will make the workforce more "balanced."

P.5: The references to the Federal Government are gratuitous. The Congress has just begun applying standards heretofore applicable to only the private sector to itself. Past Federal practice was simply a historical accident rather than a data based decision. Until recently, women were prohibited from military service in battle sectors. This position is also changing rapidly. The argument that public safety officers be treated like Federal employees is specious. There have been no studies conducted by Federal agencies with relevance to age and performance. The Department of Defense is just now initiating research related to fitness testing.

P.5-6: IAFF asserts that there are no fitness tests currently available for making retirement decisions. This is the easiest criticism to address. Virtually every public safety department in the country currently uses fitness or physical ability tests to make hiring decisions. Our survey (included in our final report) shows that over 90% of the departments use some ability/fitness test. These are exactly the same tests that would be used for making retirement decisions. We have described available tests in great detail in our final report.

The assertion that these tests are being "struck down by courts" is similarly wrong. If that were the case, no physical ability hiring tests would be used in any city. As an example, in New York City, 30,000 applicants for fire fighter positions are taking a physical ability test. Similarly, in Buffalo, New York, 3,000 applicants are taking a physical ability test for fire fighter. The Buffalo test is being given with the express permission of the United States Department of Justice and its Civil Rights Division. It is interesting to note that in both of these tests, a hose pull event plays a major role. There are literally thousands of suits in Federal District Courts that pertain to employment discrimination. Citing one such case in New Hampshire is hardly instructive when a major policy issue such as this legislation is being considered. Further, the agency responsible for enforcing employment law (the EEOC) has indicated that fitness-based retirement policies are compatible with existing federal statutes, including the ADA and the CRA of 1991. It is safe to say that in every state represented in the House Committee membership, physical ability tests are currently being used to hire public safety officers. These tests are not "being struck down" by the courts.

The IAFF argument that fitness tests do not represent the real world of public safety can also be easily countered. If they are useless as they stand, why are they currently being used by all public safety agencies? Why don't these agencies simply hire the youngest applicants by lottery? The reason is obvious. Physical ability tests provide critical information about performance capability.

P.7: The IAFF believes that there are no tests currently available that are valid predictors of age. This belief is counter to thousands of studies and to current practice in the departments that the IADD purports to represent.

P.7: The IAFF talks about the Penn State Study as my study. The IAFF uses terms like "he had pre-determined", "his research", "Dr. Landy selected studies...". As the IAFF well knows, our team consisted of more than 20 internationally known experts, working for a total of 22 months on the study. Further, our work was reviewed by an independent panel of scientists appointed by EEOC.

IAFF gives me much more credit than I deserve. I was simply a team member with some administrative duties.

The studies we "selected" number over 5,000. These studies appear on a data disk that has been circulating in the general scientific community since our literature review was completed and appear in our report as an appendix. IAFF can identify no study that was not uncovered in our research. If there were studies that refuted our results, IAFF would certainly have cited them. In addition to our exhaustive literature search, we advertised for relevant studies in all professional and scientific publications. We are certain that no relevant research escaped our attention.

P.8: The assertions about my personal and professional integrity are shameful. EEOC reviewed my work and credentials carefully before awarding the contract to Penn State. Further, EEOC appointed an oversight committee to review our work. When this charge originally surfaced during the Senate hearing of 1994, Mr. Gallegos, the Acting Chair of EEOC, responded by letter to Senator Metzenbaum and assured the Senate Subcommittee that there was no conflict and no bias as a result of EEOC oversight procedures. The results of the study must stand on their own and don't need to be interpreted in the light of any bias, real or imagined.

A poorly designed or executed study will fall of its own weight. Our study has received nothing but praise from the scientific community, those best in a position to evaluate its conclusions. The charge of Bias or conflict is unadorned mud-slinging and has no place in this debate.

CONCLUSION

The simple fact is that the IAFF wants the legislation to protect pension benefits and to avoid the possibility of fitness testing for its members. These issues have nothing to do with the scientific foundation for using fitness tests for making retirement decisions. They choose to allege inadequacies in the study to further their non-scientific agenda. This agenda happens to be at odds with statutes that protect workers from unfair discrimination based on age. If the proposed legislation is to be adopted, it must be done on grounds other than scientific since the results of the Congressionally-mandated study are unequivocal. The use of age for making retirement decisions is without scientific foundation and violates federal statutes.

EXHIBIT A

EQUAL EMPLOYMENT OPPORTUNITY COMMISSION
WASHINGTON, D.C. 20507

OFFICE OF
THE CHAIRMAN

September 22, 1993

The Honorable William D. Ford
Chairman
Committee on Education and Labor
U.S. House of Representatives
Washington, D.C. 20515

Dear Chairman Ford:

This is in response to your July 30, 1993 request for the Equal Employment Opportunity Commission's views on H.R. 2722, the "Age Discrimination in Employment Amendments of 1993".

H.R. 2722 would amend the Age Discrimination in Employment Act in two general respects. First, all state and local governments would be permitted to use age as a basis for hiring and retiring law enforcement officers and firefighters. Second, the bill would prohibit state and local governments from retiring elected judges who attain the age of compulsory retirement prior to the expiration of a judge's term of office. If the amendments are signed into law, the pertinent parts of the ADEA would read as follows:

> Sec. 4(j) It shall not be unlawful...to fail or refuse to hire or to discharge any [law enforcement officer or firefighter] because of such individual's age if such action is taken --
>
> (1) with respect to...an individual...[who]...has attained
>
> (A) the age of hiring and retirement in effect under applicable state or local law on March 3, 1983; or
>
> (B) if such age was not in effect under applicable state or local law on March 3, 1983, 55 years of age...
>
> Sec. 11(f) The term "employee" means an individual employed by any employer except that the term...shall not include any person elected to public office in any state or political subdivision...by the qualified voters thereof... The exemption set forth in the preceding sentence...shall not include with respect to retirement an elected judge before the expiration of the term of office in which such judge attains the age of compulsory retirement.

The Honorable William D. Ford
Page two

The EEOC has reviewed H.R. 2722 and we provide the following substantive and technical comments:

GENERAL COMMENTS

First, we address the law enforcement and firefighter provisions in section 4(j). As you know, the 1986 ADEA Amendments created a temporary exemption permitting age-based hiring and retirement decisions in these public safety occupations through December 31, 1993. At the same time, Congress charged the EEOC and the Department of Labor with conducting a study to determine whether tests are available that could replace the use of age as a predictor of job performance. The central research questions were:

> If one wanted to replace age with performance or capacity tests, would it be possible? Are there job-related tests that are practical, safe and cost effective? If so, can the tests be fairly administered without unduly compromising personal safety, public safety, or agency efficiency?[1]

In October 1992, this study[2] -- organized and structured by researchers from Penn State University -- was sent by the two agencies to Congress. The Study Group concluded that (1) age is a poor predictor of performance in public safety occupations, (2) practical tests are currently available that are better predictors, and (3) the temporary exemption should be permitted to expire as scheduled. H.R. 2722 rejects all of the above conclusions by, in essence, making permanent the exemption created in 1986. Moreover, Section 2(a)(B) is more expansive than the current exemption in that it permits the use of age even by state and local governments that had not implemented age limitations for public safety officers in prior years.

If signed into law, H.R. 2722 would undercut years of EEOC litigation (pre-1987) in which we routinely challenged the use of arbitrary age limitations by police and fire departments. Further, the proposed amendment to permit State and local

---

[1] Edwards, R. *Mandatory Retirement: Police, Fire Fighters and Tenured Faculty*, Public Administration Review, Vol. 53, No. 4, 1993.

[2] *Alternatives to Chronological Age in Determining Standards of Suitability for Public Safety Jobs, Volume I: Technical Report*, Center for Applied Behavioral Sciences, Pennsylvania State University, (University Park, PA: Intercollegiate Research Programs, 1992), 459 pp.

The Honorable William D. Ford
Page three

governments to require the retirement of firefighters and law enforcement officers as early as age 55 is inconsistent with a substantial body of case law which has been developed previously under the ADEA that prohibited mandatory retirement of law enforcement officers and firefighters on the basis of an arbitrary age cut-off. See e.g. EEOC v. Kentucky State Policy Dept., 860 F.2d 665 (6th Cir. 1988), cert. denied, 49 FEP Cases 1640 (1989); EEOC v. Pennsylvania State Police, 829 F.2d 392 (3rd Cir. 1987); EEOC v. Mississippi State Tax Commission, 873 F.2d 97 (5th Cir. 1989)(en banc).

Finally, we address the elected judges portion of H.R. 2722. Currently, the ADEA contains a blanket exemption for persons elected to public office by the qualified voters of a state or political subdivision of a state. H.R. 2722 would make an adjustment in this blanket exemption solely for elected judges. Under the bill, an elected judge would be entitled to complete his term of office even though attaining mandatory retirement age in advance of the full term. While it is unclear why this bill is being offered, its objective of removing one use of an arbitrary age limitation is in keeping with the spirit of the ADEA. On the other hand, the ADEA and Title VII have contained blanket exemptions for elected public officials since their dates of enactment. Thus, this bill affects a long-settled aspect of civil rights law.

TECHNICAL COMMENTS

Subsection (A) of H.R. 2722 would permit both maximum hiring ages and mandatory retirement ages. Subsection (B), to be applied when (A) is inapplicable, permits only mandatory retirement at age 55. This omission regarding maximum hiring ages may produce the curious circumstance of departments having to prove that age is a bona fide occupational qualification for hiring decisions even though they would have a specific statutory right to use age for retirement decisions.

With respect to elected judges, the bill appears to apply the ADEA only insofar as protecting an elected judge against age-based mandatory retirement prior to the completion of a full term of office. The limited nature of this objective would be made clearer were the bill to contain the word "solely" prior to "with respect to retirement."

In summary, the EEOC believes that the proposed legislation -- which would continue the exemption for State and local firefighters and law enforcement officers -- is inconsistent with Congress' purpose in 1986. Further legislation on this issue was to be based on the results of the study the Commission was

The Honorable William D. Ford
Page Four

charged by Congress to conduct. The findings contained in the Congressionally-mandated study on age and public safety jobs concluded that valid and job-related tests are viable alternatives to basing hiring and retirement decisions on age alone.

Thank you for providing the EEOC an opportunity to comment on this legislation. The Office of Management and Budget advises that, from the standpoint of the Administration's program, there is no objection to the submission of this report.

Sincerely,

Tony E. Gallegos
Chairman

88-349 O - 95 - 4

EXHIBIT B

Response to International Association of Fire Fighters Criticism of the Penn State Study on Mandatory Retirement of Public Safety Officers

April 2, 1993

Frank J. Landy, Director
Center for Applied Behavioral Sciences

Preface

In preparing this response, I have solicited comments from the various members of the Penn State team who are experts in the issues raised in the IAFF report. It is important to keep in mind that the Penn State team members represent leading researchers in cardiology, gerontology, epidemiology, exercise physiology, pulmonary function, work performance, police and fire administration, statistics, and research methodology. In addition, team members have been heavily involved in developing fitness tests and standards for police and fire departments, as well as analyzing data on ability decline and physiological capability of all age groups of the population. Our team was selected to complete the project by the Equal Employment Opportunity Commission on a competitive basis. In addition, our work plan and work products were reviewed by an independent panel of experts on aging, physiology, and medicine hired by EEOC for the express purpose of providing peer review and oversight of our work. These experts approved of our research plan, made suggestions regarding analyses, and reviewed our final conclusions and recommendations. These experts were uniformly positive about our results and concluded that we had conducted a high quality scientific investigation of the issue.

We are confident of our conclusions and recommendations and believe that the IAFF criticisms are without substance. We believe that their criticisms are directed toward policy issues and not scientific issues. Given their role as an advocacy group -- a union -- neither we nor their members would expect them to be objective or to concentrate on scientific issues. On the other hand, the Penn State team had no agenda other than a scientific one; none of us represented a group or a position.

What follows is our detailed response to the IAFF criticisms. We choose to concentrate on the IAFF report since the IACP testimony and submission was nothing more than an endorsement of the IAFF criticisms. For ease of presentation, we will respond to the issues as they appeared in the report, citing pages and paragraphs where appropriate. We strongly urge subcommittee members to review the report prior to subcommittee discussion. After reading the report, it will be clear that the IAFF claims are simply incorrect. As an example, they claim that we did not examine their data. We did. Further, they criticize us for not establishing fitness standards or recommending tests. It is clear that the congressional mandate was to determine if age was a reasonable criterion and, if not, to suggest alternatives. We present substantial evidence that age is an inappropriate criterion for making personnel decisions and identify at least one test for each critical ability. Further, we did, as Congress requested, identify the standards that tests should meet if they are to replace age. As our responses below will show, the IAFF report is highly stylized, subjective, and self-serving. It does not further the cause of either the fire fighters they represent or the public served by those fire fighters. It also does an injustice to the fire fighters who work hard to keep themselves in the condition necessary to protect the public, their co-workers, and the public from injury and death. Below, are specific responses to the statements and criticisms of the IAFF.

IAFF Introduction

1. IAFF was invited to be a member of an advisory committee. They were presented with the work plan, the anticipated products, the schedule, etc. and were asked to comment and provide any suggestions for improvement. They were also asked to facilitate the distribution of a questionnaire and to help in the identification of relevant data sets that might be analyzed in the study. There was never any intention to do anything other than brief the advisory team on the plan of action and the results of the study and ask for their help with logistics. The purpose of these briefings was to provide an opportunity for various stakeholders in the legislation to comment on areas relevant to their constituencies. We were well aware of the resistance of IAFF to the study and to any position other than a fixed retirement age (preferably a low one) that would be compatible with current pension systems and the "20 years and out" philosophy. IAFF representatives stated that in the first meeting.

2. The IAFF report states that "At no time were the IAFF data solicited, including our extensive annual death and injury surveys."

Response: That statement is not correct. I have included as **APPENDIX A** a letter from Richard Duffy dated February 25, 1991. At the subcommittee hearing (Panel III), Mr. Duffy reiterated that assertion. As you can see, that letter was written in response to our request for the IAFF survey data and the questionnaire used to solicit those data. Based on those documents, it became clear to us that the IAFF data could not be of any substantial value to us since they did not provide sufficient detail about the circumstances of illness and injury among fire fighters. This was particularly true given the data that we had already received from the National Fire Protection Association (NFPA). I have enclosed a letter that we sent to NFPA and their response to that letter as **APPENDIX B**. As you can see, we were aware of both the IAFF and NFPA data sets. The IAFF assertion that we ignored their data and similar data sets is simply wrong and the accompanying documents support our position. It is difficult to understand either the foundation or the motivation for the claim that we were unaware of their data. We made extensive use of various data sets and these data sets are well described in our technical report. It would seem that the IAFF claim was intended to make us look sloppy, biased, or both. We suggest that the IAFF review their correspondence file before repeating this baseless claim.

3. In their introduction, IAFF simply asserts that "...there is no scientific or medical evidence provided by the Penn State study that supports the conclusion that the public well being would be better served by a testing regiment (sic) than by a chronological age decision rule when it comes to retirement decisions."

Response: On the contrary, in Chapters 4 and 6 and in the numerous analyses presented in the appendices to the report, the scientific and medical evidence is presented in great detail. For whatever reason, IAFF simply disregards that material. We are at a loss to do anything other than refer to our many pages and hundreds of tables and analyses in rebuttal. In light of such substance, their argument lacks credibility.

IAFF Main Text

P. 1, Par. 2, l. 6 - Congress also excluded tenured college professors from the ADEA provisions. This is not a physically demanding job and has little in common with the job of fire fighter, police officer, or corrections officer. Thus, Congress had several parameters in mind when the exclusion was included in the 1986 amendment.

P. 3, Par. 3, ll. 2-4 - Our report explicitly deals with the interplay of age, safety, occupational demands, occupational exposure and fire fighting. In fact, it is the only study to date that does tie all of these issues together to reach a conclusion and recommendation.

There is no other document that approaches the scope and depth of our study. If IAFF were aware of such a document, they would have listed it.

P. 7, last Par. - IAFF claims that we were "unable to demonstrate that physical and mental fitness tests exist which are valid measurements of job fitness; failed to identify particular tests which could make these measurements most effectively; and refused to recommend specific standards such tests should meet."

Response: Chapter 5 reviews cognitive and fitness tests identified as measuring abilities needed to perform public safety jobs. Appendices 5-A through 5-E summarize specific information about cognitive, perceptual, personality, motor skills, and physical fitness tests reviewed. Table 5-6, on p. 5-27, summarizes our recommendations concerning tests to use to measure physical abilities. We are at a loss to understand how they could claim that we have failed to identify specific tests in the presence of the information presented in Table 5-6.

The IAFF document completely misconstrues the charge received by EEOC and transferred to us. The charge was to identify the standards that tests should satisfy, not identify standards for performance. In fact, we present a very elaborate set of rigorous standards that must be met by tests including such dimensions as reliability, job relatedness, feasibility, ease of administration, etc. These standards are compatible with standards identified in or suggested by the Uniform Guidelines for Employee Selection, the guidelines adopted by federal agencies for administration of Title VII of the 1964 and 1991 Civil Rights Acts.

PP. 7-10 of the IAFF report display quite clearly the fact that they did not understand the purpose of the study. This can be seen most clearly in the material that IAFF chooses to box on pp. 8 and 9. We agree completely that this was our charge and our report identifies the types of tests and even individual tests as well as the standards for such tests. Given the dramatic differences between IAFF and Penn State/EEOC in the understanding of the charge, it may be that all else that follows in the IAFF criticism is flawed. They simply do not understand what was to be accomplished.

P. 10 - The IAFF report criticizes us for surveying fire fighters in only 10 out of 163 departments, or .001% of fire departments. Further, they say that data collection was unsupervised, that the survey asks for "gut" reactions, and that we used average rating scores to determine essential functions.

Response: Two basic surveys were conducted (in addition to the first survey that told us which departments had what information). One survey determined essential functions (defined as tasks having consequences for public safety), the other obtained ability ratings for essential tasks. Their reference to 10 departments applies to the essential function survey. These departments are listed in the attached Table 1 (derived from Table 3-3 of the report). An additional 27 departments were sampled during administration of the ability survey. These departments were selected to be as geographically representative as possible given the time and money constraints of the project. We also attempted to get a reasonable cross-section of large and medium size departments and we believe that we were successful. In addition, team members (Landy, Fleishman, Gebhardt) had done literally hundreds of analyses of critical tasks and abilities associated with fire fighting in large and small cities throughout the country over a period of 25 years. We are well informed regarding critical tasks and abilities. Our collective experience in analyzing the job of the fire fighter was one of the reasons we were chosen to complete the study.

Also summarized in Table 1 are the numbers of fire fighters, lieutenants, captains, and battalion chiefs who completed each survey. Surveys were administered by project staff members. This was necessary in order to be sure that instructions were understood, questions were answered, and surveys were completed. The surveys were too complicated for a mail out sampling.

The decision rules applied to the essential function data are described on pp. 3-4 and 3-18. Average ratings were used, in that we calculated the average rating across raters in a given category for a given task. Reliabilities of these average ratings are given in Table 3-4 on p. 3-17. The <u>decision rule</u> that we then applied to these average ratings in order to determine essential functions is described on p. 3-18. Any task with a mean rating of 3.0 or higher for <u>any category</u> of individual affected was listed as an essential function. The cut point was given by the public consequences rating scale. A 3 or 4 on this scale indicated that injury or death was either somewhat or very likely if the task was performed incorrectly or not at all (when it should be). A 1 or 2 indicated that injury or death was either somewhat or very unlikely. Thus, a public safety task was defined as a task that might lead to someone's injury or death if performed incorrectly. Note that <u>nowhere</u> does IAFF say that our analyses of tasks or abilities was <u>wrong.</u> If there were flaws in our method, analysis, or collected knowledge base, it is incumbent on IAFF to illustrate <u>which</u> tasks were not critical and <u>which</u> abilities were unnecessary. They do not identify any such tasks or abilities. They argue not with the results but with the process.

Table 1

Summary of Survey Data Gathered by Fire Department Location

Location	Survey 1	Survey 2
Daly City, CA		4 ff, 2 capt
LA County, CA		3 ff, 2 capt
Oxnard, CA		6 ff, 1 capt, 2 ch
Sacramento, CA		2 ff, 1 capt, 1 ch
South San Francisco, CA		4 ff, 2 capt, 1 ch
Ventura, CA		1 ff, 1 lt, 1 capt
Boulder, CO	2 ff, 1 capt	
Pueblo, CO	7 ff, 1 lt, 1 capt	
Ft. Lauderdale, FL		2 ff, 2 lt, 1 ch
Hollywood, FL		5 ff, 1 lt
West Palm Beach, FL		3 ff, 2 lt
Chicago, IL	10 ff, 1 lt, 1 capt, 1 ch	
Farmington Hills, MI		3 ff, 2 lt, 1 ch
Kalamazoo, MI		6 ff, 2 lt, 1 ch
Southfield, MI		2 ff, 1 lt, 2 capt
East Orange, NJ		3 ff, 1 lt, 1 capt, 1 ch
Santa Fe, NM	8 ff, 2 lt, 1 ch	
Buffalo, NY	6 ff, 1 lt	
New York City, NY	6 ff, 1 lt, 1 ch	
Rochester, NY		4 ff, 1 ch
Fayetteville, NC		2 ff, 2 lt, 2 capt, 1 ch
Altoona, PA		4 ff, 2 capt, 1 ch
Harrisburg, PA	9 ff	
Philadelphia, PA	7 ff, 2 lt	
Austin, TX		5 ff, 1 lt, 2 capt, 1 ch
Denton, TX		2 ff, 1 capt, 1 ch
Fort Worth, TX		3 ff, 1 capt, 1 ch
Irving, TX		6 ff, 2 lt, 1 capt

Location	Survey 1	Survey 2
Plano, TX		3 ff, 1 lt, 2 capt
Chesapeake, VA		10 ff, 3 lt, 2 capt, 1 ch
Hampton, VA		4 ff, 2 lt
Newport News, VA		6 ff, 2 lt, 1 capt, 1 ch
Virginia Beach, VA		6 ff, 1 capt
Seattle, WA		4 ff, 1 capt
Tacoma, WA		6 ff
Madison, WI	11 ff, 3 lt, 1 capt	
Milwaukee, WI	6 ff, 1 capt, 1 ch	

37 Departments, 13 States	10 departments, 72 fire fighters, 11 lieutenants, 5 captains, 4 chiefs	27 departments, 109 fire fighters, 25 lieutenants 28 captains, 16 chiefs

Totals: 181 fire fighters,
36 lieutenants,
33 captains,
20 chiefs

The data that form the foundation for our conclusions and recommendations were carefully gathered from a representative sample of fire fighters and officers. The questionnaire was carefully constructed and administered. The assertion by IAFF that we collected "gut" responses and "first impressions" is incorrect. Data analyses attest to the reliability and psychometric integrity of the questionnaire responses.

P. 11 - Heart attack data presented in IAFF report.

<u>Response:</u> Without knowing the absolute numbers associated with the percentages in this table, the data are meaningless. In order to have any meaning, the base needs to be all fire fighters or the population in general. Only then can it be seen that the number of deaths being estimated for older fire fighters (in proportion to their absolute numbers) is vanishingly

small. In the NFPA report for fire fighter fatalities for 1990 (1991 and 1992 reports are not yet available), the conclusion is drawn:

"Problems still exist in two areas; heart attacks and motor vehicle accidents. More than one third of the deaths were due to heart attacks. Again this year, most of the victims with reported medical histories had prior heart problems or detectable heart conditions but continued to participate actively in fire fighting" (emphasis added).

They further state that

"There were 37 heart attack victims, 14 were shown to have prior heart problems -- usually heart attacks or bypass surgery; medical documentation showed that 5 others had severe arteriosclerotic heart disease. One other fire fighter had hypertension. No medical documentation was available for the other victims."

There were a total of 102 line of duty deaths of fire fighters reported in 1990 and one third (the 37 reported above) were heart attack victims. Virtually all of those victims with complete medical histories could have been identified as high risks through testing (but certainly not by age since they had not yet retired). Of the 102 deaths (total), 43% occurred on the fire ground (or roughly 16 deaths). We know from our survey that fewer than 2% of paid fire fighters who perform fire fighting duties are 60 or above. By extension then, we can conclude that the numbers represented by the percentages on p. 11 of the IAFF report are exceedingly small. If 102 deaths occurred in line of duty 1990 and if 37 of those deaths were heart attacks, and if 43% or 16 of those occurred on the fire ground and if 2% of active fire fighters are in the 60+ age group, the percentages offered by IAFF must represent two or three fire fighters. Further, the NFPA report suggests that even these fatalities could have been prevented through screening or testing. It is important for the subcommittee to keep in mind the absolute number of deaths that are represented by these percentages. One should have little confidence establishing public policy on the basis of one or two individuals who had detectable heart problems. It is ironic that IAFF implies that these individuals should not be tested but should be permitted to work regardless of symptomology as long as they fall below some magical chronological age. This is poor protection for the public or fellow fire fighters.

PP. 11-12 - With respect to the central question of the ability of older fire fighters to perform their jobs, the IAFF criticism misses the point. They argue that injuries and health problems increase with age, and that the emotional and physical stressors of fire fighting may take a unique toll (pp. 11-12). They also argue that the average fitness level in the general population is misleading for fire fighters.

All those points, however, are consistent with the position that functional and health status should determine continued employment, not age alone. Without conducting a specific evaluation of fire fighters, the available data indicate the likelihood that at least some older individuals probably can continue doing the job quite well. (It is important to keep in mind the EEOC threshold for a BFOQ, i.e. all or almost all people of that age or above must be incapable of performing the critical function; in practice, this is usually interpreted as 95%.) The argument in the fire fighters' response is that people cannot do the job because of injuries and health problems, and that since these are associated with age, then age should be used to exclude people. It seems more appropriate to say that when people can no longer perform at certain levels, then they should retire, whatever their age.

The section in the IAFF report on aging and abilities clearly overstates the implications of cognitive and sensory changes with age. In most areas, age differences at age

50 or 60 are small and have little or no practical implications. Differences are found in standardized laboratory settings which tend to maximize these differences, because they do not allow individuals to draw upon experience in solving problems. Experience and judgment contributes to maintaining performance, even when primary mental abilities show some decline. The material cited concerning shifting attention (p. 14) is picked out of a vacuum. Those studies involve the ability to shift in boring, repetitive situations, not akin to fire fighting. Moreover, the differences at age 60, for example, are not going to be large enough to have much practical value. The section about sensory abilities was simply incomprehensible. We are aware of no findings that would suggest such uniform decrements as implied in the IAFF report, with such sweeping practical implications.

An overall consideration is the distinction between average abilities of an age group and individual performance. A central finding in research on aging is that individual differences grow larger over time. On most measures, many older people (e.g. 60) will score as well as younger people. Performance will be maintained especially by people who practice abilities. For example, someone who exercises daily in a vigorous way will remain fit--not at the peak of a 20 year old athlete, but quite fit, nonetheless. Whether that older person can do a specific task (e.g., fire fighting) cannot be determined by age alone.

P. 12, ll. 1-4 - As indicated earlier, we were aware of and used the NFPA data. It is cited in both Chapters 4 and 6 of our report. In addition, our report cites multiple additional data sets on both the general population and public safety officers. Our experts used the "actuarial" data to come to conclusions regarding risk of catastrophic medical events while performing public safety tasks. In fact, our conclusions are based on thousands of cases represented in those data sets. IAFF would have the reader believe that we conducted our work in a data vacuum, supported only by ill informed and biased views of the job of a fire fighter. On the contrary, we know the job quite well and our conclusions are firmly based on empirical data and relevant analyses of large scale data bases.

P. 12, l. 10. - The Penn State study is criticized for "insist(ing) on assessing age in a vacuum."

Response: This was the purpose of the study, i.e. to deal with age by itself rather than as confounded by illness, disease or lifestyle. We consider this to be the strength of the study rather than a weakness. This is exactly the point of ADEA, the validity of age as a BFOQ.

P. 13, l. 1 - The IAFF document asserts that it is "axiomatic that cognitive and physical abilities weaken with age."

Response: There is nothing axiomatic about the relationship at all. The entire knowledge base of gerontology and life span development contradicts that assertion. Crystallized intelligence, tacit knowledge, and judgment can all be seen to increase with age rather than weaken. Further, our data show very clearly that there is no regular decline in either cognitive or physical abilities that would apply to all people. In fact, there are wide individual differences in decline and these differences become more pronounced as one ages, making age an increasingly poor predictor of capacity or performance. The IAFF simply repeats a pernicious stereotype and labels it "axiomatic."

P. 15, Par. 1 - The critique that the 30-40% reduction in maximal cardiac output from age 20 to age 70 years affects the body's ability to process oxygen, which is needed to sustain prolonged physical exertion, is erroneous. In fact, for a given submaximal aerobic workload, the amount of oxygen required to carry out the work is unaffected by age. The only time the age related reduction in maximal cardiac output would affect a fire fighter's ability to work aerobically is if the cardiac output demand to meet the oxygen uptake of the task at hand exceeds the individual's maximal cardiac output. Therefore, to implicate cardiac output and its relationship to age as a major factor in determining work performance is fallacious.

In all the work on self-contained breathing apparatus, no research documents differences in cardiovascular stress (or cardiac output) not accounted for by the extra weight of using an SCBA.

The IAFF concerns about heat stress are an extrapolation of laboratory findings to a workplace situation where the learned work practices of the worker are such that life threatening thermoregulatory problems cannot be linked to age.

The IAFF report raises a concern about age-related decrements in pulmonary function and the exacerbation of these decrements with the use of an SCBA. Raven (one of the Penn State team members) has continuously been on record (published in the scientific literature) that this is not the case; in fact in some instances, it has been predicted that the SCBA will <u>enhance</u> lung function.

PP. 16-17 - The IAFF report singles out the Sothmann study as evidence that all or almost all fire fighters over the age of 65 cannot meet minimal standards of aerobic fitness.

<u>**Response:**</u> On pp. 4-45 through 4-48 of our report, we show very clearly that the extent to which fire fighters can meet minimal standards is substantially affected by the fitness programs that exist within their departments. The data shows that fitness programs can dramatically affect whether a 60 year old can or cannot meet standards. This means that the issue is not age but exercise.

P. 17, l. 1 - The IAFF report argues that age is a fairer method for making retirement decisions.

Response: This implies that age is a perfect predictor of performance. We have shown convincingly in our study that this is not the case. In fact, an age-based retirement rule is manifestly unfair to the person over the mandatory age who can still perform the job.

P. 18, l. 10 - The IAFF report asserts that fire fighters are healthier than the population in general "due to screening at the hiring stage for cardiovascular disease."

Response: IAFF has inadvertently endorsed the findings and recommendations of our study. If people can be screened at hiring stages they can be screened at any point in their career. By default, IAFF agrees that cardiovascular screening is both feasible and valuable. We agree. That was the point of our report. The screening mechanism is valid and less discriminatory than age.

P. 19, Sec. 5, Par. 1

Response: The critical public safety tasks require the ranked physical abilities listed on pages 3-85 to 3-90 of the Penn State report. Tests of these abilities are routinely used in the general population to assess fitness and have been extensively used for entry level testing of potential employees. Such tests are listed on page 5-27 (Table 5-6) of the Penn State report. There is no reason to assume that the nature of the relationship between physical abilities and performance of public safety tasks would be markedly different in older individuals. Indeed, a good example of this is heat tolerance. Everyone would agree that the job of fire fighting is unique in that they have to work in excessive heat. There is also evidence that older individuals have reduced heat tolerance. However, as we state in our report (p. 4-31), ... VO_2 max is "the primary first-order individual predictor of heat tolerance. Age played no primary or secondary role in predicting heart rate, core temperature, or rate of heat storage." The Penn State team considered the unique environmental conditions under which public safety employees work and then considered whether age or physical ability measures would be better predictors of human performance. In no case was age shown to be markedly unique. The two studies showing relationships between physical abilities and fire fighter performance up to age 50 years (highlighted by the IAFF) bear out the limitations of using age as a predictor of specific physical tasks performed by fire fighters.

P. 20, ll. 7-11 - The IAFF report presumes that standards cannot be set by municipalities.

<u>Response</u>: The fact that we did not recommend minimum standards does not mean that it cannot be done. This was not our responsibility. Our study was a feasibility study, not a standards study. In fact, municipalities set standards now. Almost 90% of the fire departments that responded to our survey reported that they had physical performance standards for hiring purposes. Thus, it is clear that departments can set standards. In fact, several team members (Landy, Fleishman, Gebhardt, Sothmann, Kenney) have frequently assisted fire departments in identifying and setting performance standards. But general standards cannot be set that ignore departmental differences or job differences. Standards can and must be locally determined if they are to meet the rigorous review of the <u>Uniform Guidelines on Employee Selection</u> followed by the EEOC in determining job relatedness.

P. 20, Aging and Strength - IAFF cites data related to decline by decade as evidence of the relationship between strength and age.

<u>Response</u>: In the first paragraph (p. 20), IAFF selected information related to the decline in upper body endurance (muscular endurance) and lower body strength by decade. Although these figures were found when testing incumbent fire fighters, the IAFF failed to include the remainder of the research that showed a small decline across decades in strength

when public safety personnel participate in a fitness program. These data are further supported with research by the National Institute on Aging that found a stabilization and some increases in strength in 50 to 80 year olds who participated in a fitness program.

The Training and Age section of the Aging and Public Safety report (p. 4-63) demonstrated that there is an increase in the cross section of the muscles for the older individual who participates in a strength training program. Cady et al. showed an increase of 12% in work capacity for fire fighters age 50 or greater who participated in a fitness program. Similarly, Gebhardt et al. showed that the decline in strength per decade was reduced to 1-3% for men and 2-7% for women when a fitness program was used by public safety personnel.

The Aging and Public Safety report demonstrated that participation in a strength training program can reduce the decline in strength with age. The fire fighter job is physically demanding and does not require daily performance of arduous tasks that would help the individual maintain a fitness level required to perform the job effectively. Therefore, it is incumbent upon fire fighters to find another means to maintain a level of fitness that allows them to perform the essential job functions. This can be accomplished with participation in a fitness program that includes strength building exercises. The point is that age does not predict decline in strength but disuse does. Functional tests would tell us a great deal more about strength than would age.

The IAFF report also addressed the decline in motor ability with age and singled out fine motor ability using an example of coupling hoses. Fine motor ability is defined as small manipulations such as buttoning a shirt. The coupling of hoses is a more gross movement once the couplings have been matched. A study that investigated incumbent fire fighter and applicant performance on a variety of fire fighting tasks found that applicants could couple the hoses faster than incumbents. However, the incumbents were not found to be deficient in this task. Therefore the use of hose coupling as a fine motor task is inappropriate. The IAFF report authors do not understand the technical nuances of the terms they use. As a result, they choose incorrect data and studies to make their points.

P. 22, l. 12 - IAFF asserts that we have demonstrated that age is a BFOQ.

Response: We have done exactly the opposite. We have shown that there is no one age that might be set above which all or almost all fire fighters would be unable to perform the job. If such an age were to be set, it would be very high, possibly in the high 70's. The easiest way for IAFF to contradict this assertion is to suggest a specific age that would serve as a BFOQ. It is clear that no such age exists and that virtually any age in the range from 55 to 70 would be arbitrary and would fail to meet the standard set by ADEA, i.e. all or almost all of incumbents of that age or above would not be able to perform the job in question. It is generally accepted that "all or almost all" would translate to 95% of the people of a given age. Our data show clearly that age, construed in this manner, cannot be a BFOQ. It is interesting to note that IAFF is willing to let local municipalities determine what age should be set for mandatory retirement but demand a national and general standard if tests are to be used. Their position is inconsistent.

To put the IAFF argument in simpler terms, it is clear that there is an age beyond which all of us will be dead (e.g., 120). Nevertheless, there will be tremendous differences in the ages at which each one of us dies as well as how well we are functioning when that death occurs.

It is our belief that there is unlikely to be any substantial compromise of performance as a result of age alone until the age 70 or beyond. And even at that age, there is likely to be sufficient variation such that one could not conclude that all or almost all of the incumbents of that age could not do the job.

PP. 23-25 - IAFF assertion of statistical flaws.

Response: In regard to the "statistical flaws" alluded to in the Fire Fighters' brief (PP. 23-25), it should be pointed out that incidence of cardiac events is nearly always greater among individuals with negative test results than among those with positive tests results, because of the much larger proportion of individuals with negative tests. This is a well-established epidemiologic principle that is being manipulated to support the Fire Fighters' brief. The bottom line is that we are not supporting the value of testing to identify individuals at risk for sudden incapacitation in asymptomatic individuals. There is little to be gained with respect to protecting the public. The medical event that will compromise public safety (i.e. heart attack or stroke) is exceedingly rare and unlikely.

To provide a simple example of the flawed analyses and logic of the IAFF argument, consider the AIDS population. Following IAFF logic, the population with AIDS would be healthier than the rest of us even if 100% of this population died because 100% of a small population (i.e., AIDS population) would yield a smaller absolute number of deaths than occur in the remainder of the general population.

P. 26, l. 2 - IAFF asserts that age is a fairly applied criterion.

Response: As indicated above, we do not agree. It is certainly not fairly applied to the individual capable of continued high level performance who is forced to retire.

P. 37, Para. 2 - The IAFF report asserts that tests are "less uniform," "more discriminatory," and "inherent(ly) imprecis(e)."

Response: This is simply not true since it implies that age is a uniform, non-discriminatory and precise predictor of performance. The only prediction that one can make from age is how long ago you were born. Age is a great deal less uniform, more discriminatory and less precise than tests for assessing functional capacity.

Concluding Comments

Penn State University conducted a scientific study to determine the extent to which age could be used to predict job performance for public safety officers. The multidisciplinary team considered the issue from a wide variety of perspectives, including medical, physiological, gerontological, psychological, legal, and industrial. We reviewed over 4000 scientific publications and re-analyzed numerous data sets. The team concluded that age was a poor predictor of performance for public safety officers and that other, more precise and less discriminatory methods were available for making retirement decisions. The team had no preconceived position and had no stake in the outcome of the study.

In contrast, IAFF is a union rather than a scientific organization. They are a clear stakeholder in the outcome of the study. Having fought hard for many years to establish pension rights for their members at an early age, they are not enthusiastic about any study that suggests that fire fighters can work well past current retirement age. In addition, ADEA makes it impossible to identify an age beyond which testing might occur since this would represent prima facie evidence of discrimination (i.e. disparate treatment). As a result, if our recommendation is accepted, departments will have a mechanism for introducing fitness standards and testing for all ages. This, of course, would serve the public good and public safety, but individual fire fighters who are not capable of meeting those standards would be threatened by such a procedure. IAFF would be expected to defend members against such testing. It is hard to understand their logic on this point since the public safety is not served by allowing individuals at risk for a catastrophic medical event (e.g. heart attack) who are below mandatory retirement age to actively engage in fire fighting, while preventing older fire fighters with no such risk from similar activities. Burying one's head in the sand does not make the problem go away.

The argument that age is a good predictor of performance is untenable. Similarly, the argument that testing is not sufficiently developed to take the place of age as a decision mechanism for retirement is also demonstrably incorrect. The report outlining the scientific rationale for the elimination of the public safety exemption is exhaustive and represents the state-of-the-art in this area. The arguments of IAFF have been shown to be specious and should be disregarded in the debate on the scientific merits of the age issue. To be sure, there may be policy issues that argue in favor of the exemption, but there is no scientific foundation for continuing that exemption.

APPENDIX A

INTERNATIONAL ASSOCIATION OF FIRE FIGHTERS

ALFRED K. WHITEHEAD
President

VINCENT J. BOLLON
Secretary-Treasurer

February 25, 1991

Bob Zance
Penn State University
CABS
205 Research Building D
University Park, PA 16802

Dear Bob:

Thank you for your recent request concerning fire fighter safety and health. In accordance with your request, we have enclosed the following information:

Last five D&I Surveys
1989 Data Questionnaire

If you need further assistance, please let us know.

Sincerely,

Richard M. Duffy
Director
Department of Occupational Health and Safety

RMD

opeiu #2, afl-cio
file#OSHA- 1991-0215

1750 NEW YORK AVENUE, N.W., WASHINGTON, D.C. 20006 • (202) 737-8484 • FAX (202) 737-8418

APPENDIX B

617-984-7461
Try Allison w/ ?? if questions
week of 9/9

PENNSTATE

1855

(814) 863-1011
FAX: (814) 865-3309

Center for Applied Behavioral Sciences (CABS)
Aging and Public Safety Project

The Pennsylvania State University
207 Research Building D
University Park, PA 16802

MEMO

617-770-3000
617-984-7429

To: Rita Fahy, NFPA
From: Bob Vance, 814-865-3313, 814-865-3309 Fax
Subject: Request for Reports and Analyses
Date: August 30, 1991

As we discussed on Wednesday, I am passing along a request from Debbie Gebhardt for injury and illness information on firefighters. Listed below is the information she has requested. Will you provide the reports, and conduct any analyses of your data base as needed in order answer these questions, assuming that the relevant data aren't already in the reports? Please send me a bill for the costs of these reports and analyses. If you have any questions, please call me at 814-865-3313 or Debbie Gebhardt at 301-699-0770. Thanks very much.

1. Copies of the following reports:

   a. US Fire Fighter Deaths (for the three most recent years)
   b. US Fire Fighter Injuries (for the three most recent years).
   c. US Fire Department Profile Through 1989
   d. Fire Fighter Casualty Study Package (1984-1988)
   e. Selected Special Analyses of US Fire Fighter Fatalities (1987-1990)
   f. Fire Fighter Casualties as a Result of Roof or Floor Collapses in Wood-Frame Buildings (1990)

2. Questions a, b, and c pertain to the 1989 Fire Fighter Deaths report (or possibly the "Contract Report on the 1990 Fatalities"). If the information needed to answer these questions is not in the Report, can the database be analyzed in order to answer them?

   a. the absolute number of deaths by age category as stated in Figure 6
   b. the absolute number of deaths per age group
   c. for those individuals who died, can we determine how many were known to have heart disease, past injuries, or other medical problems?

3. Questions a, b, c pertain to the 1989 Firefighter Injuries Report. If the information needed to answer these questions is not in the Report, can the database be analyzed in order to answer them?

   a. for individuals who suffered injuries, can we determine how many were known to have heart disease, other past injuries, or other medical problems?
   b. can injury data be broken down by age categories and whether the firefighter was career or voluntary?
   c. what illnesses and injuries have been analyzed in relation to the population protected (look at Tables 1, 2, and 3, by type of injury)?

NFPA®

National Fire Protection Association

International

Executive Offices
1 Batterymarch Park
P.O. Box 9101
Quincy, Massachusetts 02269-9101 USA
Telephone (617) 770-3000
Telex 200250 Fax (617) 770-0700

September 4, 1991

Mr. Bob Vance
Center for Applied Behavioral Sciences
207 Research Building D
Pennsylvania State University
University Park PA 16802

Dear Bob:

Enclosed are the reports that you ordered and the invoice. The attached table displays the information that I believe Question 2 of your letter addresses. Because we update our files continually, the totals reported on these tables will not exactly match what was published in the article. If I've misunderstood the questions, please let me know and I'll try again.

As far as Question 3 goes, we do not have the detail from our survey of fire departments to answer parts (a) and (b). Mike Karter, the author of the fire fighter injury study, could perform the custom analysis necessary to find the answers to part (c). The minimum fee for a custom analysis is $495 for members and should cover the staff and computer time needed to complete the analysis. If you want the analysis done, you should call him directly to discuss details about what exactly it can and should cover. His number is (617) 984-7465.

If you have any questions about any of this material, I'll be back in the office September 16. If you need assistance before then, you can try Alison Miller at 984-7461. She may be able to track down any additional material you need.

Sincerely,

Rita F. Fahy, Manager
Fire Data Bases and Systems
Fire Analysis and Research Division

Enclosures

cc: Mike Karter
Alison Miller

Fire Fighter Deaths 1985 - 1989

Age Groups	Total Deaths	Heart Attack Deaths
Under 20	19	3
20 - 29	90	4
30 - 39	121	23
40 - 49	152	85
50 - 59	142	105
60 and Over	102	72

Fire Fighter Deaths 1989

Age Intervals	Total Deaths	Heart Attack Deaths
20 and Under	5	1
21 - 25	8	0
26 - 30	10	2
31 - 35	14	6
36 - 40	8	4
41 - 45	14	6
46 - 50	11	9
51 - 55	12	11
56 - 60	16	10
Over 60	19	14

Fire Fighter Deaths 1989
by Physical Condition

Condition*	Total Deaths	Heart Attack Deaths
Illness, prior health problem**	20	19
Diabetes	4	4
Severe arteriosclerotic heart disease	9	8
Hypertension	1	1
Other	1	1

* Physical condition prior to fatal injury is frequently not reported.
** This is the category where prior heart attacks and bypass surgery would be coded.

Testimony Regarding Mandatory Retirement and Chronological Age Presented to
Committee on Education and Labor Subcommittee on Select Education and Civil Rights

March 24, 1993

Dr. Frank J. Landy
Director: Center for Applied Behavioral Sciences
Penn State University

I am pleased to have the opportunity to speak with you today about chronological age and job performance. In a 1986 amendment to the Age Discrimination in Employment Act of 1967, Congress prohibited employers from forcing workers to retire when they reached a particular age unless the employer could show that all or almost all of the people of that age could not perform the essential functions of the job in question. But as a result of the critical role of public safety officers--i.e. police officers, fire fighters, and corrections officers--in protecting the lives and property of citizens, Congress specifically excluded this occupational group from the provisions of that 1986 amendment. They also excluded tenured college teachers but others will speak to that issue. I am here to discuss retirement of public safety officers. This exclusion for public safety officers was granted for a limited period - until December 31, 1993. The Department of Labor and the Equal Employment Opportunity Commission were given the charge of conducting a study to be done that would provide a scientific foundation for deciding whether to make the exclusion permanent or eliminate the exclusion, thus treating public safety officers like all other workers for purposes of the Age Discrimination in Employment Act. Penn State University was chosen to complete that study. I directed the Penn State research team and I will summarize our findings and the method we used to arrive at our conclusions and recommendations.

The team that was assembled to carry out the research included leading experts in areas such as work performance, gerontology, occupational medicine, exercise physiology, epidemiology, statistics, law enforcement and fire administration, and public policy. Our work spanned 22 months and included travel to many of the major cities in the country to collect information from fire fighters, police officers, and correction officers. In addition, hundreds more cities, counties and states completed extensive questionnaires about the nature of public safety jobs in their agencies. As examples, if we just look at the states represented by members of this subcommittee, we received assistance from 9 agencies in New York, 6 agencies in New Jersey, 3 agencies in Montana, 19 agencies in Virginia, 13 agencies in Ohio, 8 agencies in North Carolina, 2 agencies in Nebraska, and 14 agencies in Illinois. Finally, we received and analyzed data sets related to illness, injury, and fitness of public safety officers from various federal, state, county, and municipal agencies.

As a research team, we met regularly to plan data collection and data analysis as well as consider the results and implications of those analyses. By the time we were finished, we had examined data from almost 500 of the largest public safety agencies across the United States. In addition, we met with representatives of federal agencies, management organizations, and labor unions. Our final report runs to several thousand pages, including extensive literature reviews, data analyses, and legal reviews. These were the activities that provided the foundation for the conclusions and recommendations I will present to you this morning.

The ultimate question that we had to answer was as follows: Is there a particular age beyond which public safety officers actually become a liability or a jeopardy to the public they are sworn to protect? In other words, can we tell simply by knowing someone's age that they will perform poorly in their job? The 1986 amendment to the ADEA has already settled that issue for almost all American workers but not for public safety officers.

In considering age and the performance of public safety officers, there were two separate issues to consider. One was the probability of a catastrophic medical event (for example, a heart attack) occurring while a critical public safety task was being performed (e.g., rescuing a person from a burning building), thus resulting in injury or loss of life to the citizen. The second issue related to the gradual deterioration of abilities due to age. The question here was whether or not a specific age could be identified beyond which abilities necessary for the efficient performance of public safety tasks were radically diminished, again leading to the possible injury or loss of life for citizens. A final question we were to consider was related to alternatives to chronological age for making retirement decisions. If age is not a good criterion for retirement decisions, is there any feasible alternative?

Before answering those questions, let me clarify the issue of age and performance as we considered it. We are considering the role of age by itself. As we get older, our lifestyle changes, our nutritional habits change, and we may begin to experience the cumulative effects of illnesses and injuries. Thus, many of us feel that as we get older, we become less fit. But we can change our diet, begin to exercise more regularly, and get more sleep and we notice that those effects of aging go away. What this really means is that they were not the effects of aging at all. Our study deals with the effects of age per se and separates that effect from the effects of lifestyle, illness, and injury.

Now, to our conclusions and recommendations. First, it is our conclusion that age is a very imprecise predictor of who will experience a catastrophic medical event such as a stroke or heart attack, particularly if the individual in question has no prior history of heart disease or current symptoms. There are many accurate procedures available that have the advantage of increased precision without the discriminating effect of labeling all people of a particular age as somehow incapable of performing their duties or as unacceptable risks to the public safety.

Similarly, we believe that age is a poor predictor of ability decline. By ability, I mean things like stamina, strength, reasoning, and comprehension. While it is generally true that there appears to be a gradual decline of abilities with the aging process, there are still substantial variations within any age group in terms of ability level. This means that some 60 year old workers are actually higher in ability level than their 40 year old counterparts. Further, experience appears to make a substantial contribution to effective work performance and the relationship between age and experience is obvious.

Finally, we came to the conclusion that there are tests currently available that could be used to monitor the fitness of public safety officers of any age. In our report, we identify such tests.

These conclusions, then led to our recommendation. We recommend that the current exemption to the ADEA which permits public safety employers to require retirement at a particular age be eliminated and that public safety officers be treated like all other American workers. We do not believe that the public safety will be jeopardized in any way by allowing this exemption to expire on December 31, 1993, as the law is currently written.

It is clear that age is a cheap and easily administered "test." Any alternative will be more expensive and challenging to administer. Nevertheless, the question we must ask is whether the current practice, permitting employers to force retirement based on age, is fair to the older worker or results in greater safety to the public. In our mind, the answer to both questions is "no." With respect to alternatives, we recommend tests. Employers use tests now to hire these public safety officers, to promote them, and to maintain fitness standards. There is no reason why these tests and similar tests cannot be used to administer or inform retirement policies. So in some senses, the feasibility question has already been answered. For that reason, it is our belief that continuing to use age as a criterion for making retirement decisions would be unfairly discriminatory and would not contribute to the safety of the public.

I would be pleased to answer any questions you might have about our research, conclusions, or recommendations. In addition, I would be pleased to comment on any policy issues that you feel might be relevant.

April 2, 1993

Congressman Major R. Owens,
New York, Chairman
Committee on Education and Labor
Subcommittee on Select
Education and Civil Rights
U.S. House of Representatives
518 House Annex I
Washington, DC 20515-6107

Dear Congressman Owens:

I was pleased to be able to address the Subcommittee on Select Education and Civil Rights on March 24, 1993. As you know, at that meeting I became aware of the criticisms of the International Association of Fire Fighters and the International Chiefs of Police. Since the IACP criticisms simply repeated the IAFF points, I will deal with the IAFF report in some detail. Since many of the IAFF criticisms appeared to be technical, thus attempting to undermine the conclusions and recommendations of the Penn State study, I felt that the Subcommittee members should consider our response before proceeding with their deliberations. I appreciate the opportunity to submit this response and would be pleased to talk with you or other subcommittee members should you have any questions. I assume from your comments at the hearing that our response will be entered in full into the hearing record.

Cordially,

Frank J. Landy, Director
Center for Applied Behavioral Sciences

FJL/jls

Enclosures

Testimony Before the Committee on Labor and Human
Resources, Subcommittee on Labor
Regarding Mandatory Retirement and Chronological Age

April 19, 1994

Dr. Frank J. Landy
Director: Center for Applied Behavioral Sciences
Penn State University

I am pleased to have the opportunity to speak with you today. I directed the study mandated by Congress to examine the feasibility of substituting tests for chronological age in making retirement decisions regarding public safety officers.

I led a Penn State University research team that included internationally known scientific experts in areas such as work performance, gerontology, occupational medicine, exercise physiology, epidemiology, statistics, law enforcement and fire administration, and public policy. In addition, most of the members of our research team have been conducting research on the physical demands of law enforcement and fire fighting for several decades. Thus, our team was scientifically and practically familiar with the demands of the jobs in question.

Our work spanned 22 months. Over 500 of the largest cities, counties and states completed extensive questionnaires about the demands of public safety jobs in their agencies, the injury and death rates in their departments, and their retirement policies.

We also analyzed data sets related to illness, injury, and fitness of public safety officers from various federal, state, county, and municipal agencies. We used that information to complete an exhaustive job description of both the tasks carried out by public safety officers as well as the knowledges, abilities, and skills necessary to carry those tasks out successfully.

Our final report runs to several thousand pages and carefully considers the results of over 5000 research studies. It is the largest and most comprehensive scientific study of these issues ever to have been conducted and has been given high marks for breadth and accuracy by knowledgeable scientists and researchers both in the U.S. and abroad.

The ultimate question that we had to answer was as follows: Can we tell simply by knowing someone's age that they will perform well or poorly in their job? The 1986 ADEA amendment has already settled that issue for almost all American workers. The only issue that remained was to answer that question specifically for public safety officers.

In our study, we considered two primary issues related to performance -- incapacitation due to illness and ability decline. The most important question we were to answer was whether tests could substitute for age in making retirement decisions.

88-349 O - 95 - 5

Before answering those questions, let me clarify the issue of age and performance as we considered it. We are considering the role of age by itself. An individual's physical condition, and subsequent job performance, can be affected by many variables including lifestyle, nutrition, exercise, and commonly experienced illness. Many older workers maintain a healthy life style and nutrition, follow a program of exercise, and take preventive actions to avoid illness, and, as a result, are successful in their work long after younger colleagues begin to decline. Neither the effective work of the older employee or the ineffective work of the younger employee is the result of age per se. Instead, the work effectiveness reflects non-age related variables. Our study focused on the effect of age per se on performance and separated that effect from the effect of lifestyle, illness, or injury.

Now, to our conclusions and recommendations. First, it is our conclusion that age is a poor predictor of who will experience a debilitating illness at the work site, particularly if the individual in question has no prior history of illness or current symptoms. There are many accurate procedures available for identifying underlying illness that have the advantage of increased precision without the discriminating effect of labeling all people of a particular age as somehow "prone" to one or another illness.

Similarly, we believe that age is a poor predictor of current ability. By ability, I mean things like stamina, strength, reasoning, and comprehension. The age-related decline of job-related abilities is slight. In addition, there are substantial variations within any age group in terms of ability. There are 30 year olds who cannot climb a stairway pulling a charged fire hose and there are 60 year olds who can. Further, it is clear that experience makes a substantial contribution to effective work performance. Younger public safety officers invariably turn to their older colleagues to take the lead in difficult and dangerous actions -- and those older colleagues invariably oblige.

Finally, we came to the conclusion that tests are a safe and effective alternative to age in making retirement decisions. The best example to support that conclusion is provided by police and fire departments themselves. These departments are currently using tests for hiring that could be just as easily used for making retirement decisions and in our report, we identify and describe available and, in our opinion, legally defensible tests for the important job-related abilities required of public safety officers. By using such tests, a police or fire department will also be in compliance with the Civil Rights Act of 1991 and the Americans with Disabilities Act.

These conclusions, then led to our recommendation. We recommended that the exemption to ADEA be eliminated and that public safety officers be protected from age discrimination like all other American workers. Since that exemption expired on December 31, 1993, current policy is in line with our recommendation. It is our belief not only that the public safety is not jeopardized by the elimination of that exemption, but also that the increase in the experience base and the protection of the rights of additional workers actually adds to the public good. The scientific basis for this conclusion is presented in our report. From the scientific perspective, the proposed legislation (HR 2722) would be a step backward.

I would be pleased to answer any questions you might have about our research, conclusions, or recommendations and their implications.

EXHIBIT C

Massachusetts State Police
Commissioned Officers Association

April 18, 1994

Senator Edward M. Kennedy
Senate Labor and Human Resources Committee
Washington, D.C. 20510

Dear Senator Kennedy,

As president of the Massachusetts State Police Commissioned Officers Association, I am writing to you to express our opposition to HR 2722.

We strongly feel that state police officers over the age of fifty-five can perform all of the duties incumbent upon them, as long as they are physically able and willing, therefore we are in support of our association members who meet that criteria and support them in their efforts.

Sincerely,

Joseph C. Saccardo
President

EXHIBIT D

Massachusetts Police Association

OVER 17,000 MEMBERS

April 18, 1994

Hon. Edward M. Kennedy
Chairman
Senate Committee on Labor
and Human Resources
315 Russell Senate Office Building
Washington, D.C. 02510

Dear Senator Kennedy:

We understand that H.R. 2722 is scheduled for hearings before your committee on April 19, 1994. On behalf of our 17,000 law enforcement members throughout Massachusetts, we respectfully urge you to do whatever it takes to defeat this bill and protect the ability of our members to continue working.

This bill would permanently exempt Massachusetts and other states from complying with the Age Discrimination in Employment Act, at least with respect to some police officers and fire service personnel, and could result in a substantial number of our members being forced to retire before their time.

The defeat of H.R. 2722 would leave in place current retirement laws, but would permit a police officer, or a group of police officers to challenge a mandatory retirement age. We believe firmly that the ability of any public employee to challenge a mandatory retirement must be protected as a basic tenet of democracy and a fundamental right to guard against age discrimination.

We offer whatever help you deem necessary in this matter, and we thank you for your help.

EDWARD M. MERRICK
Executive Board
Legislative Director

Legislative Office • One Beacon Street, Suite 1120 • Boston, Massachusetts 02108 • (617) 720-3477/(800) 445-0911 FAX: (617) 720-4

2300 M STREET, N.W., SUITE 910
WASHINGTON, DC 20037
PHONE: (202) 466-7820
FAX: (202) 466-7826

POLICE EXECUTIVE
RESEARCH FORUM

EXHIBIT E

CHUCK WEXLER
EXECUTIVE DIRECTOR

Police Executive Research Forum Position Statement

Public Safety Exemption to the Age Discrimination in Employment Act (ADEA)

March 29, 1994

Background

In 1986, Congress passed an amendment to the ADEA that eliminated mandatory retirement for all employees. Law enforcement officers and fire fighters were granted a seven-year exemption from the law, so that a study could be done to determine whether age should be a bona fide occupational qualification for public safety positions. The exemption expired on January 1, 1994.

The study, conducted for the Equal Employment Opportunity Commission by Pennsylvania State University, concluded that the chances of an asymptomatic officer experiencing a catastrophic medical event while carrying out his or her responsibilities is extremely small and not well-predicted by age. The study also found that age-associated declines in cognitive and physical abilities that could potentially affect an officer's capacity to carry out public safety tasks are highly modifiable, can be "attributed to injury, illness and lifestyle variables," and vary widely from one individual to another. The study further concluded that there are tests available for predicting the probability of both catastrophic medical events and age-associated declines "that are more effective than age and have lesser adverse impact on older employees." Based on these and other conclusions, the study recommended that the public safety exemption be eliminated.

Position

Prior to the expiration of the public safety exemption, PERF surveyed all members to determine an organizational position on this issue. Seventy-three percent of responding PERF members supported performance-based retirement (opposed an extension of the police exemption); 19% supported mandatory retirement for police (supported an extension of the public safety exemption); and 8% indicated PERF should not take a position on this issue. PERF members also debated the merits of the public safety exemption at their 1993 annual meeting; many of the participants in the discussion seemed to favor performance-based retirement.

(over)

Based on the membership survey results, the annual meeting discussion and the Penn State study conclusions, the Police Executive Research Forum supports performance-based retirement policies for public safety employees; in addition, PERF opposes a reinstatement of the public safety exemption to the ADEA. It is our belief that public safety employment decisions should be made solely on the basis of a person's ability to perform the essential functions of a job, rather than non-performance based criteria such as age, disability, race, sex and other similar characteristics protected by federal, state and local law. For additional information, please contact Karin Schmerler of PERF at (202) 466-7820.

Promoting 16 Yea[rs] Progressive P[olicing]

FACT SHEET

POLICE EXECUTIVE RESEARCH FORUM

The Police Executive Research Forum (PERF) is a national membership organization of progressive police executives from the largest city, county and state law enforcement agencies. PERF is dedicated to improving policing and advancing professionalism through research and involvement in public policy debate. PERF's primary source of operating revenues are government grants and contracts and partnerships with private foundations and other organizations.

BACKGROUND: Before the mid-70s, most police executives accepted the status quo in law enforcement and rarely discussed issues among themselves. But, as crime increased and new kinds of information became available through research, police executives began to question traditional police practices. In 1975, ten police executives from some of the nation's largest cities met informally to discuss common policing concerns. Driven by the successful exchanges that took place at their meeting, the chiefs decided to meet on a regular basis to explore tough issues and develop a professional body of knowledge aimed at reducing crime and improving the quality of policing.

PERF GOALS: Incorporated in 1977, PERF was founded to:

- Improve the delivery of police services and crime control nationwide.
- Encourage debate of police and criminal justice issues within the law enforcement community.
- Implement and promote the use of law enforcement research.
- Provide national leadership, technical assistance and vital management services to police agencies.

MEMBERSHIP: PERF membership requirements are unique among law enforcement organizations in that potential members must be nominated and approved by current PERF general members. To qualify for general membership, a police executive must hold *a four-year degree from an accredited institution and head a department of at least 100 full-time employees or a jurisdiction of at least 50,000 persons.* They must also be *committed to the organization's principles and willing to take on tough issues.* To obtain different viewpoints, which enable PERF to formulate innovative responses, PERF offers other categories of membership to academicians and law enforcement personnel from smaller jurisdictions.

STAFF: PERF staff includes former police executives, criminal justice experts and professionals with extensive experience in management, research, training and policy analysis. Daily contact with police chiefs and law enforcement agencies enables PERF staff to provide answers to issues that plague police executives, such as domestic abuse, firearm-related violence, drug trafficking and the plight of the homeless.

RESEARCH: Working with police practitioners to apply research to real-life situations and tailor experiments to meet the needs of their communities is the cornerstone of PERF. PERF researches the most compelling issues facing police — how to control crime and drugs, maintain law and order and operate an efficient organization. PERF also encourages research on controversial and sensitive issues, such as the police response to the homeless and police handling of persons with AIDS.

Current PERF projects address youth gangs, police compliance with the Americans with Disabilities Act, ways of preventing and combatting urban decay and crime, police awareness of cultural differences issues, school violence, factors that influence the growth and decline of drug markets and police use of excessive force. In the past, PERF has been involved in the development and implementation of a problem-oriented approach to drug enforcement, new approaches for coping with repeat offenders, national accreditation standards for law enforcement and the testing of criminal investigation models.

INFORMATION CLEARINGHOUSE AND RESOURCES CENTER: The escalating demands for police assistance in the 1980s and 1990s by communities and political leaders require law enforcement executives to search continually for information that can help them formulate effective police programs. To communicate the latest in police-related innovation, research, legislation and other relevant issues, PERF publishes three newsletters and writes and disseminates books and reports that are widely used by police departments, colleges, the media and governments throughout the world. Frequently, PERF members are asked to serve on national commissions and testify before the U.S. Congress on important national issues. PERF staff and members are interviewed regularly by the press on matters affecting communities across the nation. They also release position papers on high-priority public safety legislation.

In 1987, PERF established METAPOL, the nationwide electronic telecommunications network through which police executives and others concerned with law enforcement issues can communicate and receive immediate feedback. Through its Management Services Program, PERF offers technical assistance, training and a wide array of services to member and non-member police organizations of all sizes. PERF also provides assistance in policy-making to law enforcement agencies through its Policy Center, which contains policy and procedures manuals collected from law enforcement agencies throughout the United States and Canada.

SENIOR MANAGEMENT INSTITUTE FOR POLICE (SMIP): SMIP is an intensive three-week course which merges real-life police experience with management techniques drawn from business school curricula. Using the case-study method, professors from leading academic institutions teach current and future police executives how to apply business principles and concepts to policing strategies.

EXHIBIT F

International Association of Fire Fighters

Brook L. Beesley
~~State Area Representative~~
16th District
P.O. Box 521, Alameda, CA 94501
Telephone (510) 638-6263 / FAX (510) 483-1306

29 June 1993

Alfred K. Whitehead
General President
International Association of
Fire Fighters, AFL-CIO, CLC
1750 New York Avenue, N.W.
Washington, D.C. 20006-5395

Dear Mr. Whitehead:

This letter responds to your letter of 24 June 1993, which permanently removes me as a State Representative for the 16th District. I have worked hard out of loyalty and respect for over seven years to promote the labor movement . . . it seems painfully obvious that the loyalty and respect will not be reciprocated by your office.

It's a struggle, at times, to stay on common ground with the International on every issue, but your blatant exercise in censorship and fear of criticism, thus another point of view, and in this case my <u>opinion</u> on the firefighters' "entry age" exemption of the Age Discrimination in Employment Act ("ADEA") is not bold leadership in my book. In addition, you letter implies that I support changing the mandatory retirement age of Federal firefighters. Check the record, I am of the personal opinion that mandatory retirement age-based policy appear to be a valid BFOQ. Still, I feel that the IAFF should compromise on this issue with Congress however. I could lobby this position better than anyone. Nonetheless, your assault cast me as an obstructionist of "political correctness," whereby ideas and speech are stifled for <u>fear</u> that they could offend certain groups, individuals or Congress. You want to play hardball with my free speech rights? I am ready to suit up.

First, your actions to permanently remove me as a State Representative is a direct hit to the 16th District and is unquestionably politically motivated. Because, as you state: "I cannot designate someone as an IAFF State Representative who disagrees so fundamentally with an important <u>policy mandate</u> (ADEA issue) by the IAFF Executive Board and three IAFF Conventions..." If you're looking for puppets as IAFF representatives I'm certainly not your man. The problem is, however, <u>nobody</u> technically services federal firefighters better than I do, period. My track record and servicing expertise stands for itself. I wasn't recommended to be a State Representative the <u>first</u> day Vice President Mike Crouse met me because I don't know what I'm doing. To the contrary, it is well known in the federal sector that I am technically skilled and can get through the federal sea of red ink, in an aggressive fashion, if necessary, on all issues I commit myself to; if you knew me you would know this to be true. I challenge you to prove me wrong.

Federal firefighters are the ones that will pay the price -- again. The situation is exacerbated, moreover, by the dismal and sluggish effort of the IAFF to provide quality "direct service" to federal firefighter issues. Federal firefighters need results, Mr. Whitehead, not rhetoric. There must be accountability at all levels. Your office should be held responsible for your performance.

Your letter isn't brave. It's self-centered, pompous and abrupt bully pulpit to control diverse ideas and political views.

Second, your unspecified letter of 26 April 1993, to rescind (temporarily) my appointment and effectively "hanging me out to dry" as an IAFF State Representative was received by my office on 29 May 1993, from Vice President Mike Crouse via fax, [and] from IAFF Legal Counsel, Joseph R. Reyna via fax on 30 May 1993. I hardly think I have refused to accept your correspondence and/or position. In fact, I have officially acknowledged your 26 April letter with Mike Crouse and Joe Reyna...

In terms of your first paragraph of your 26 April letter, there is no specificity as to any reasoning for rescinding my appointment as State Rep whatsoever. Perhaps it was because of the political pressure of the unfounded claims and allegations of raiding initiated by the American Federation of Government Employees ("AFGE"), who also, in my opinion, has a dismal record when it comes to federal firefighter issues. Clearly, in your view, I am guilty before proven innocent! That doesn't set well with me.

In terms of your second paragraph, same letter, I have always notified Mike Crouse, and copied (cc'd) all correspondence with my letterhead and more importantly, the "capacity" I was acting to his office for review, i.e., President F-259, Designated Rep for another local, or State Rep. Unless Mike wants to hang me out to dry too, and deny it, I have always been loyal in this sense with Mike and his office as 16th District Vice President.

Simply put, the nature of the evidence, not the seriousness of charges, is how all union leaders should base their decisions. In other words, I am uninfluenced by anything (e.g., AFGE raiding allegations) other than the material facts and evidence. You need to understand that leadership is not the same thing as control.

Third, you have further advised me, as stated: "I want to make it perfectly clear that you in no way represent the IAFF, the IAFF's positions, or any member of the IAFF, except your own local F-259..." Be advised (a) federal employees are entitled to designate employees of their own choosing to act in the capacity of a representative, and (b) to interfere with, restrain, or coerce any employee is unlawful. On this basis, I am currently the designated representative for IAFF, Local F-15, F-32, F-102, F-213, and F-268. I would appreciate in the future that your office cease dictating to me who I can provide representation and service thereof.

Freedom to write and to read as we please is the cornerstone of our democratic way of life; your letter shows little faith in the independent judgment of people. Censorship represents the authoritarian view that certain individuals should decide what is best for other people and, therefore, to think and believe, in this case the ADEA issue.

My sense is now, anyone foolhardy (especially from the insignificant 16th District) enough to question, much less contradict the your office will be subjected to the heavy hand of the IAFF Executive Board. Remember, you are a servant of the IAFF, like any union leader, and are employed to promote service to our membership.

Sincerely yours,

//s//

BROOK L. BEESLEY
President
Local F-259, IAFF

cc: Vice President
16th District, IAFF

Director
Government Affairs, IAFF

Committee on Education and Labor
(Subcommittee on Select Education and Civil Rights)

Interested Congressional Representatives
U.S. House/Senate

Superintendent, 5th VP
International Association of Chiefs of Police

Deputy Executive Director
Virginia Municipal League
National League of Cities

INTERNATIONAL ASSOCIATION OF FIRE FIGHTERS

ALFRED K. WHITEHEAD
General President

VINCENT J. BOLLON
General Secretary-Treasurer

June 24, 1993

BY REGISTERED MAIL

Brook L. Beesley
1312 Pierce Ave.
San Leandro, CA 94577-2442

Dear Brother Beesley:

On April 26, 1993 I sent you a letter rescinding your appointment as an IAFF State Representative for District No. 16. In that letter I stated that I would be revisiting this issue within six months and would be making a further determination at that time. However, in view of recent actions by you, I am permanently removing you from the list of individuals who may be designated a State Representative for District No. 16.

Your recent letter to 16th District Vice President Michael Crouse, which we received in our offices on June 9th, directly challenges the IAFF's carefully developed position on continuing the fire fighters' exemption from the maximum hiring age and mandatory retirement age requirements of the Age Discrimination in Employment Act. Although we can disagree on the merits of this issue, your letter was sent on letterhead indicating that you are a "State Area Representative" for the 16th District. You had already been suspended by me at the time you wrote the letter, but you continued to misrepresent your status with the International.

What concerns me is that this letter has been copied (cc'd) to the House Subcommittee on Select Education and Civil Rights, which is currently considering our proposed legislation, and which has already heard IAFF testimony on this issue. The letter was also sent to the International Association of Chiefs of Police, and the National League of Cities, who have supported us on this matter.

In my opinion, this letter may be misconstrued as coming from the IAFF because you sent it over letterhead indicating that you are an IAFF State Area Representative.

I cannot designate someone as an IAFF State Representative who disagrees so fundamentally with an important policy mandated by the IAFF Executive Board and three IAFF Conventions. The continuation of the ADEA exemption is the subject of a major legislative effort by the IAFF and is supported by the ADA/ADEA Committee, which is composed of an International Vice President and eight (8) representatives of major IAFF municipal locals or state associations.

In addition, the arguments advanced in your letter could be used by certain members of Congress to support changing the current mandatory retirement age for Federal fire fighters, which is 55 years old. This would disrupt the current retirement system governing Federal fire fighters and could mean they would have to work longer. Your arguments certainly give our opponents ammunition in this direction.

Although you may continue to represent your clients in the manner you see fit, you may not call yourself an IAFF State Representative or "State Area Representative" under any circumstances. I must also emphasize that I am making every Federal local in the IAFF aware of my decision only because in the past you've refused to accept my

1750 NEW YORK AVENUE, N.W., WASHINGTON, D.C. 20006-5395 • (202) 737-8484 • FAX (202) 737-8418

Brook L. Beesley
June 24, 1993
Page - 2

correspondence which imposed your suspension. I want to make it perfectly clear that you in no way represent the IAFF, the IAFF's positions, or any member of the IAFF, except your own local, F-259 - Alameda Naval Air Station.

Sincerely,

Alfred K. Whitehead
Alfred K. Whitehead
General President

AKW/JRR/bv
cc: Vincent J. Bollon, General Secretary-Treasurer
IAFF Executive Board
IAFF Federal Locals
ADA/ADEA Committee

INTERNATIONAL ASSOCIATION OF FIRE FIGHTERS
FAX TRANSMISSION SHEET

TO: Brook Beesley, President

ORGANIZATION: IAFF Local F-259

FAX NUMBER: 510-483-1306

DATE: 4-30-93 TIME: 3:20 p.m.

FROM: Joseph R. Reyna, Legal Counsel

PHONE NUMBER: (202) 737-8484

FAX NUMBER: (202) 737-8418

NOTES/COMMENTS:

NUMBER OF PAGES INCLUDING COVER SHEET: 2

INTERNATIONAL ASSOCIATION OF FIRE FIGHTERS
1750 NEW YORK AVENUE, NW
WASHINGTON, DC 20006-5395

Joe

INTERNATIONAL ASSOCIATION OF FIRE FIGHTERS

ALFRED K. WHITEHEAD
President

VINCENT J. BOLLON
Secretary-Treasurer

April 26, 1993

Brook L. Beesley, President
IAFF Local F-259
P. O. Box 521
Alameda, CA 94501

Dear Brother Beesley:

This letter is to advise you that, effective immediately, I am rescinding your appointment as an IAFF State Representative. You are no longer authorized and are specifically prohibited from carrying out any function and/or duty of the International Association of Fire Fighters.

Further, it has come to my attention that you have in the past been working independently of your 16th District Vice President, Mike Crouse. You should be aware that any and all assignments and approval for them are required to be processed through your District Vice President.

I shall be revisiting this issue within six months and will be making a further determination at that time.

Sincerely and fraternally,

Alfred K. Whitehead

Alfred K. Whitehead
General President

cc: Michael J. Crouse, Vice President, 16th District

LAW OFFICES OF
CHRISTOPHER WOLCOTT KATZENBACH
MILLS BUILDING - PENTHOUSE ONE
220 MONTGOMERY STREET
SAN FRANCISCO, CALIFORNIA 94104
TELEPHONE (415) 391-6464
FAX (415) 391-2698

CHRISTOPHER W. KATZENBACH
KERRY R. STOEBNER

DAVID B. HARRISON
OF COUNSEL

May 5, 1993

Alfred K. Whitehead
General President
International Association of
Fire Fighters
1750 New York Avenue, N.W.
Washington, D.C. 20006

Re: Brook Beesley

Dear Mr. Whitehead:

I have been asked to reply to your April 26, 1993, letter to Mr. Brook L. Beesley, President of IAFF Local F-259, which rescinded his appointment as IAFF State Representative.

Your action appears to be in response to certain unfounded claims by the American Federation of Government Employees (AFGE) that Mr. Beesley and his local are "raiding" fire fighters currently represented by the AFGE. It is disturbing, to say the least, that mere allegations of "raiding" by the AFGE could lead to Mr. Beesley's removal as State Representative.

By precipitously removing Mr. Beesley as State Representative, you have given credibility to the AFGE's allegations of "raiding". This action harms Mr. Beesley's stature as a union representative and indicates that the IAFF will not support Mr. Beesley when push comes to shove. The federal agencies Mr. Beesley deals with are quick to exploit any sign of weakness and will exploit this situation to harm the interests of fire fighters in the various locals Mr. Beesley has been assisting since becoming State Representative.

The weakness of the AFGE's claims against Mr. Beesley is illustrated by its citation to Local F-259's newsletter. The criticisms of the AFGE attributed to this newsletter came in response to an attempt <u>by the AFGE</u> to raid Local F-32 and get those employees to switch to the AFGE from the IAFF. It is unbelievable that the AFGE would use its own raid on the IAFF in support of a claim that the IAFF is raiding the AFGE. Surely the IAFF has the right to resist such a raid by urging its members to remain with the IAFF and to criticize the AFGE's raiding tactics.

We are confident that the Article XX proceedings initiated by the AFGE will clear the air on this "raiding" issue and demonstrate once and for all that there is absolutely nothing to the AFGE's claims.

However, unless you reinstate Mr. Beesley as State Representative, the damage to his stature and ability to represent fire fighters will remain whatever the outcome of the Article XX proceedings. Mr. Beesley's loss of stature will directly impact his credibility on major issues, including the threatened base closures in the Bay Area. Mr. Beesley has worked hard for the IAFF and the interests of fire fighters. He deserves and is entitled to your support.

Very truly yours,

Christopher W. Katzenbach

– Response Newsletter –

IAFF - Big Words/Small Talk:

We don't hear a lot from municipal fire departments about the way they are represented by IAFF but do we ever hear from federal locals!

The chief complaint revolves around the little effort expended in our behalf because of the extreme reluctance of IAFF to challenge the federal government. *On anything!*

When F-259 rejoined IAFF, after a nauseating period with the Federal Firefighters Association, it was because of a belief we had a better chance of representation with IAFF. FFA was formed because no one else seemed to have a real interest in the welfare of the federal firefighter. The experiment was a failure conceived by ex IAFF personnel. Some were well meaning and sincere but the major offices were held by men of questionable integrity with predicable result. End of story.

So, we came back with one firm promise to ourselves. We intended to insure IAFF would not ignore federal locals and would be of assistance.

Things started very well with 16th District VP Mike Crouse providing excellent guidance and on the spot help in contract negotiations.

From that highpoint, things went downhill; FAST! We just didn't realize it until the AMERTEK issue reared its ugly head and Local F-259 found itself in the forefront of the effort to get the Navy, and the military in general, to stop using admittedly dangerous crash/fire vehicles.

Exhibiting a typical reluctance to do the right thing, the Navy continued insisting the AMERTEK was safe while at the same time acknowledging its extreme danger to firefighters. This contradictory position *caused them no embarrassment or consternation.* Confusing to say the least for the naive members of F-259 who generally believed the Navy had some consideration for firefighters health and safety. Those of us who have a somewhat cynical outlook were not greatly surprised but did admit to disappointment. This was the beginning of F-259s struggle to force the Navy to do "the right thing" and will undoubtedly end up in federal court with big money, payola, fraud, national and international politics forced into the spotlight.

North American Free Trade considerations are most conspicuous when you remember the AMERTEK is manufactured by the Canadian Government in a sweetheart deal with the U.S. Government.

Basically the "sweetheart deal" is this; Canada manufactures second rate shoddy equipment and the U.S. buys it and swears the equipment is great. Quietly, the U.S. tries to arrange correction of a myriad of *dangerous*, and not so dangerous, deficiencies to be paid for, over and above purchase cost, by the taxpayer.

What does Canada get out of this? Money and jobs and the sure knowledge their neighbor to the south can be had. This must be reassuring.

The U.S. gets the shame and embarrassment of knowingly endangering life and ignoring criminal activity for the sake of politics.

The taxpaying public gets the middle digit.

Local F-259 figured a bit of assistance from the International would be nice so approaches were made. Guess what? IAFF tried to make F-259 back down and also tried to muzzle *RESPONSE!* articles!

We were *endangering* their Navy contacts! This came with the impact of ice water in the face. *IAFF cared more for their contacts than the safety of the federal firefighter!* Local F-259 was doing exactly what Al Whitehead advised us to do. Whitehead said in a letter to F-259 members for publication in *RESPONSE!* "The International Union is also there to help your local when times get tough. ...We will not give up until we prevail." This was in reference to a staffing problem at Lemoore Naval Air Station that IAFF worsened

Volume #2, Issue #4

and when it got tough, the IAFF sold out the local membership. To this day "Whitehead" and "IAFF" are curse words at Lemoore N.A.S.

He continued "You can rest assured that if similar misfortune hits your local, you will get the same kind of response from us." Whitehead also said in the same letter "You have a right to fair pay and decent working conditions. ...unless you actively exercise those rights, they will not do you much good." In reference to *RESPONSE!* he said "This newsletter is just the kind of forward movement out members have come to expect and deserve. Keep up the good work."

Well, Al, we believed you and went full bore with the AMERTEK issue and you tried to sell us out for the sake of "good relations" with the Navy.

Thanks, Al, we really needed that!

Local F-259 pushed and pushed and pressure on Whitehead from the Navy increased. He decided it was time to take over the AMERTEK issue and put out a letter saying he was on our side. To begin action, the IAFF first needed good information.

IAFF initiated a "survey" of AMERTEK problems. The survey asked for information not related to the AMERTEK; also information not available to locals. Not much time was allocated either. IAFF also insisted upon a signed waver of each locals interest in the problem. In other words, if you want IAFF help you must cede your rights and control of the issue to the International!

A lot of time passed and IAFF announced that the survey was nonproductive. There was nothing IAFF could do. Dead issue!

Reading the survey could lead one to suspect it wasn't designed to gather data but to wrest control of the AMERTEK issue from concerned locals and allow it to die an unnatural death.

Who is IAFF with? Us or them?

At the same time as the IAFF announcement, the General Accounting Office was looking into the issue and saying they found reason to suspect fraud and were looking further.

Recently, someone applied pressure to the GAO and now they aren't talking and refuse to share information. AMERTEK embarrasses someone and is a political hot potato related to the North American Free Trade Act.

Listen up, boys. Local F-259 will not back down. Local F-259 will not be satisfied until the crooks are punished, the Navy obeys the law and the Federal Government "does the right thing!"

Politics and back scratching are not acceptable, now or ever. Do the "right thing."

Obey the Law!

In the March/April issue of INTERNATIONAL FIRE FIGHTER, Whitehead says "As President ...I will not relinquish my responsibility to save fire fighters' lives ...but I do have an obligation to each and every IAFF member to protect them from being unnecessarily injured or killed." He continues "...the day I can't stand up for the health and safety needs of our members is the day I'll quit."

Really, Al?

Truly, Al?

Bye, Al.

The only one not heard from.

Yet!

IAFF Enforcer

International Association of Fire Fighters

Brook L. Beesley
~~State Area Representative~~
16th District
P.O. Box 521, Alameda, CA 94501
Telephone (510) 638-6263 / FAX (510) 483-1306

07 July 1993

House Education and Labor Committee
The Honorable Chairman William Ford
U. S. House of Representatives
Washington, D.C. 20515

Dear Chairman Ford:

Recently I requested from your office by way of letter dated 14 June 1993 (with enclosure dated 07 June 1993), your written position regarding the age-based public safety "exemption" from the Age Discrimination in Employment Act (ADEA), as amended in 1986.

As you can see, per enclosure, it is my belief that the actions being taken against me are in direct retaliation for my criticisms and free speech opinion (id., 29 C.F.R. § 458.2 (a)(2)) of the IAFF's position on the ADEA age-based employment criteria. I would like to clear-up an IAFF concern, however. That is, the IAFF is concerned that my ADEA position or status and/or non-status as an IAFF State Representative may be misconstrued as coming from the IAFF on this ADEA issue. The long and short of it is: The IAFF was well aware I was not on assignment for this (see 07 June 93, position paper) ADEA issue, and did not represent myself as acting on behalf of the IAFF. I was in the position of an independent designated representative for Local F-213 only, and made myself perfectly clear, as stated in my 07 June position paper: "This is simply my opinion and recommendations, "not endorsements..."

Since I have essentially been constrained by the IAFF to defend myself and what I believe in, I am forced to take up more of your valuable time or that of your staff to further explain my personal values as it relates to this ADEA exemption issue.

[F]undamentally, I have always opposed as a labor leader all forms of discrimination in the employment place and have successfully brought legal action against the U.S. Navy in the Northern District of California for their discriminatory policies respecting Height/Weight charts and standards. See *Federal Firefighters v. Boennighansen, et al.* Case No. C-88-5185-WHO (1989).

The Age Discrimination in Employment Act of 1967 follows on the heals of the Civil Rights Act of 1964; two very important pieces of legislation that have set the stage for non-discrimination policies. To my knowledge there is no actual finding or factual basis that chronological age per se is a solid predictor of abilities or performance. If this is true, Congress is entrusted, by the people and for the people, to do the right thing and uphold the concepts of non-discrimination in this great and equitable country.

The Honorable Chairman William Ford
U. S. House of Representatives
-Page Two-

And, particularly in the fire service whereas firefighters' health and safety is my main concern, keeping of course fully consistent with OPM and EEOC regulations and instructions. For example, requirements for job-relatedness of physical performance because of the obvious arduous, hazardous, and strenuous physical activity of firefighting requires a certain level of "health" and "fitness" status because of a high degree of responsibility toward the public; for federal firefighters, national security; and more importantly to the firefighter and his/her loved ones. Congress must take the lead and seize the moment on this issue and require full compliance with the ADEA and adopt certain and specific Medical and Physical standards for all firefighters (see e.g., 5 C.F.R. §§ 339.202 and 203; 54 Federal Register 9761-66). The current legislative ADEA age-based employment exemption defines far too loosely what the law allows.

The continued reliance on age-based policy that quite frankly has not made the hazardous environment on the fire ground any more safer must be reevaluated as to its effectiveness and purpose. Congress should pursue the "causes" of firefighter death and injury. Increasing firefighter survivability explicitly requires a certain level of medical and physical fitness. It's just that simple.

What's the answer? To start with, mandatory physical fitness training for the specific purpose of maintaining physical conditioning as an employment standard and business necessity. Second, and just as important, annual medical surveillance programs to safeguard the health of firefighters whose work subject us and others to significant health risks due to occupational or environmental exposure and physically strenuous activity demands. It should go without question, that any truly concerned employer and fire department would monitor their firefighters in this sense.

Demonstrating the ability to perform the actual job requirements, and developing task demonstrations and performance testing is critical to identify individuals who can safely and efficiently perform the tasks of the job. It's too late to find this information out on the fire ground when the consequences could be fatal. Such tests are "demonstrably" a reasonable measure of job performance and measures the firefighters for the job, not the firefighter and his/her age in the abstract. Validation studies (e.g., Penn State Study) over a period of many years show tests to be a good predictor of training performance and post-training success. The tests may not be job specific, but they are good predictors of a firefighter's ability to do the job physically, and have been supported by validation studies. See also, *John W. Herrington, et al. v. Williams Cleghorn, et al.*, U.S. District Court of Nevada, 1986, p. 4., and *John Herrington, et al. v. William Cleghorn, et al*, pp. 9-10 (9th Cir., 1987).

In my view, firefighters should also be entitled through proper bilateral negotiations with management, included with any physical performance testing program, compensation to offset the cost of fitness training, compensation for off-duty and on-duty training time, remedial programs for

The Honorable Chairman William Ford
U. S. House of Representatives
-Page Three-

firefighters who fail fitness tests and alternative employment for senior firefighters who are unable to continue to meet fitness qualifications. The bottom line is, the evidence shows overwhelmingly that a fit firefighter (1) performs more effectively, and (2) more safely.

In conclusion, I feel personally that our International union should emphasize the vital relevance of medical and physical requirements related to employability -- consistent with non-discriminatory policy. In addition, as a consequence of the natural progression of physical deterioration of aging, as a result of the normal aging process, and in response to the hazardous work environment of firefighting which may contribute to physical deterioration -- mandatory retirement age-based policy must be maintained.

I welcome your questions and/or that of your staff and would be honored to further explain my position. Please feel free to contact my office directly.

Sincerely yours,

BROOK L. BEESLEY
President
Local F-259, IAFF

Enclosure
cc: Vice President
16th District, IAFF

Executive Board
Local F-259, IAFF

Legal Counsel
Local F-259, IAFF

AFSCME®

JAN 27 1995

American Federation of State, County and Municipal Employees, AFL-CIO
1625 L Street, N.W., Washington, D.C. 20036-5687
Telephone (202) 429-1000
Telex 89-2376
Facsimile (202) 429-1293
TDD (202) 659-0446

January 26, 1995

The Honorable Harris Fawell, Chairman
Employer-Employee Relations Subcommittee of the
Committee on Economic and Educational Opportunities
U.S. House of Representatives
Washington, D.C. 20515

Dear Mr. Chairman:

On behalf of 1.3 million members of the American Federation of State, County and Municipal Employees (AFSCME), I am pleased to write in support of legislation which would make permanent the Public Safety exemption to the Age Discrimination in Employment Act (ADEA). I respectfully request this letter be submitted into the record.

The Age Discrimination in Employment Act prohibits the use of mandatory retirement and maximum entry ages, unless an employer can prove in court that age is a valid employment criteria. During the early 1980's, a series of lawsuits were filed challenging age limits in police and fire departments. Many of these lawsuits caused much confusion for state and local governments as some courts ruled the age limits are legitimate, while others ruled they are not. Most of the cases, however, required several years of work and hundreds of thousands of dollars in legal fees and came to no conclusion at all.

To help resolve this confusion, and to save states and localities from diverting any more of their scarce resources to litigation, Congress passed a law in 1986 which temporarily exempted public safety officers from the ADEA. The law called for a seven year exemption, during which time states and localities could continue enforcing their age limits. During this seven year period, a study would be performed to determine once and for all if age is a valid criteria for public safety occupations.

The study found that age has a direct impact on a person's ability to work as a public safety officer. Specifically, the study found that age decreases aerobic capacity and strength, and increases the risk of sudden incapacitation (i.e. heart attacks). However, because the study was unable to determine an exact age at which all people will be unfit for duty, it concluded that age is not a good predictor of ability.

in the public service

A recent statistical analysis conducted by AFSCME on normal retirement age and service requirements for correctional officers (CO's) in 20 states showed the following: Of the surveyed states, three (Connecticut, Maryland and Massachusetts) provide for normal retirement following 20 years service regardless of age. Louisiana CO's hired prior to 1986 are entitled to retire with 20 years of service regardless of age but those hired after 1986 must be at least 50 years of age.

Arizona, Florida, New Mexico and New York CO's are eligible for normal retirement with 25 years of service regardless of age. Minnesota and Wisconsin CO'S are eligible for normal retirement at age 55 regardless of years of service. All the remaining surveyed states require a minimum age and years of service in order to qualify for normal retirement.

The current legislation introduced by Chairman Harris Fawell (R-IL) would make permanent a temporary waiver from the ADEA which would allow public safety agencies the option to establish mandatory retirement and/or maximum entry ages. The bill would allow state and local governments to continue using successful age-based hiring and retirement policies for corrections and police officers, firefighters and certain other law enforcement personnel as a way to ensure an effective and balanced public safety work force.

Other reasons that we support the permanent exemption:

- Public safety officers have physically demanding jobs.
- The federal government uses mandatory retirement ages for its public safety officers, and states and local governments should have the same option to set retirement ages for their public safety officers.
- The public safety exemption is supported by those who are directly affected by it. The public safety officers who could theoretically be discriminated against if they are exempt from the ADEA are the very people seeking the exemption.
- Mandatory retirement ages have proven to be fair and effective.
- If the current public safety exemption is allowed to expire, state and local governments will be inundated with costly litigation in their effort to defend their age policies.
- Fitness tests are not a viable alternative to age limits because they do not adequately simulate the real world public safety operations; in fact, they discriminate against women and minorities; they can be dangerous; and, they are very expensive to administer. In the absence of valid fitness tests, public safety agencies have no alternative to age limits to ensure a qualified and balanced workforce.

AFSCME and its members who are public safety officers strongly support this bill. Such legislation will serve to protect citizens by helping to ensure that public safety officers are able to do their jobs. We urge the members of this subcommittee to report the bill to the full committee for consideration. Thank you.

Sincerely,

Charles M. Loveless
Director of Legislation

CML:cyb

FEB 1 1995

FRANK WALL
10453 Del Norte
Los Alamitos, CA 90720

February 10, 1995

The Honorable Harris Fawell
Chairman
Subcommittee on Employer-Employee Relations
Committee on Economic and Educational Opportunities
United States House of Representatives
2181 Rayburn House Office Building
Washington, DC 20515

Dear Chairman Fawell:

It was an honor for me to represent AARP and testify before your subcommittee on January 24, 1995.

I left the hearing quite troubled and concerned with what I felt was a lot of misleading half-truths and innuendos that was presented by Mayor Ticer, Mr. Scully and Mr. Nesbit. It is my belief that their testimonies were nothing more than a smoke screen to mask the real reason for wanting the ADEA exemption for public safety officials extended permanently.

Supposedly they feel that older police officers and firefighters jeopardize public safety. If this were the case, why is it they are opposed to testing and physical fitness programs? I'll tell you why. They do not want to subject our membership to any kind of testing.

Why? Because they know that a large number of younger firefighters and police officers may not pass the test.

They think it is okay to have younger firefighters and police officers on the job who are obese and out of shape, jeopardizing public safety.

I believe the study conducted at Penn State University, that was sanctioned by the EEOC, concluded that testing was the best way to determine a public safety officer's ability to do the job, not age. No study that I am aware of has concluded that any public safety official is more dangerous in society because he is older.

The IAFF states that there are no valid tests that can measure firefighter's ability to do his or her job. The Penn State study disagrees and I disagree. The county of Los Angeles Fire Department requires all personnel to take a periodic medical examination. The examination includes a strength test, flexibility test, blood tests, an EKG at rest - EKG while under stress on the treadmill - EKG during recovery. During the treadmill stress and recovery period, blood pressure is monitored every two minutes. The attending physician gives us a head to toe examination that is followed by a counseling session with a clinical exercise physiologist. (I am enclosing a copy of my last medical exam that will provide you with an example of what can be done in the area of testing).

Our program has come under scrutiny in the past. However, it has not been eliminated, just revised and improved.

In regard to entrance level testing, Mr. Nesbit stated that the court threw out the hose pull test in the state of Massachusetts because the females could not pass it.

We have a similar test in place as part of our entry level physical agility examination.

Our test was also challenged by certain groups several years ago. May I add, not only females protested. Our department worked to revise the test, but not to abolish it. By allowing more time for candidates to complete the test, we were successful in maintaining that portion of the exam in our entry level examination process.

Another interesting figure that Mr. Nesbit threw out was that one-seventh of our IAFF membership is over fifty years of age. That would amount to about 28,571 members. He then stated these men accounted for one-third of all firefighter injuries and deaths in the line of duty. I find this very difficult to accept. I recall three very recent fire tragedies: Seattle, Washington; Altadena, California; and Gleenwood Springs, Colorado. These fire tragedies resulted in twenty firefighters losing their lives. Not one was over fifty years of age.

As I stated at the hearing, at the South Canyon fire in Colorado (the tragedy on Storm King Mountain), there were 49 firefighters on the fire line. The oldest was thirty-two years of age. Fourteen firefighters died in this fire. Nobody can say for sure what might have happened if some older, seasoned firegighters with more experience would have been on the mountain that day. Maybe it deserves some serious consideration.

What really convinced me that Mr. Nesbit's testimony lacked any credibility what so ever, was his statement that our IAFF Local Union 1014 in the County of Los Angeles overwhelmingly voted in favor of mandatory retirement.

This is just not true. In fact it is a blatant lie! Our 2500 plus membership never had the opportunity to participate in a vote. There was no vote. I know, I am a member of Local 1014. I never received a ballot nor was I ever asked my opinion. I have not been able to find one member that voted or was asked how they felt on this important issue.

I am sure you recall Mr. Nesbit's answer to Congressman

Martinez's question on how the vote was conducted and the results tallied. Mr. Nesbit stated that he thought the vote was conducted at the union hall during a Board of Directors meeting. That is not true. I have several friends on our Board of Directors, they told me the mandatory retirement issue had been discussed at some Board meetings, however no vote was taken. All eleven board members are well aware of the IAFF position on the issue. IAFF representatives in California have been relieved of their union roles for voicing disagreement on this issue. If all eleven Directors voted in favor of mandatory retirement, that is a far cry from an overwhelmingly majority of the 2500 plus membership.

Mr. Scully stated that all police organizations in the nation unanimously favor mandatory retirement. This is not true either. For example, the 17,000 member Massachusetts State Troopers Association and the Police Executive Research Forum oppose it.

AARP and the EEOC have received hundreds of letters and calls opposing mandatory retirement.

After leaving your hearing on January 24, I felt I was the only person in the world that was not in favor of mandatory retirement. I have been involved with this issue on a personal basis since 1992. That is when I joined AARP and started following the events leading to your committee meeting on January 24, 1995. I truthfully can not say if I am a minority on this issue. I know that there are many throughout the nation that feel as strong as I do. I have talked to lots of people in all walks of life that feel an individual should be allowed to work as long as they can perform their duties.

I would like to thank you for allowing me the opportunity to present our side of the story. I enjoyed discussing the few issues that were brought up at the hearing. Before any bill moves out of your committee, I hope serious consideration is given to all the possible ramifications the legislation would have on public safety, plus the health and welfare of all police officers and firefighters.

Please make this letter a part of the record of the hearing.

Sincerely,

Frank Wall

Frank Wall

cc: The Honorable Matthew G. Martinez
United States House of Representatives
2231 Rayburn House Office Building
Washington, DC 20515

